Powers of Nature

Prepared by the
Special Publications Division

National Geographic Society
Washington, D. C.

POWERS OF NATURE

WILLIAM R. GRAY, TEE LOFTIN, TOM MELHAM,
CYNTHIA RUSS RAMSAY, JUDITH E. RINARD,
Contributing Authors

Published by
THE NATIONAL GEOGRAPHIC SOCIETY
ROBERT E. DOYLE, *President*
MELVIN M. PAYNE, *Chairman of the Board*
GILBERT M. GROSVENOR, *Editor*
MELVILLE BELL GROSVENOR, *Editor Emeritus*

Prepared by
THE SPECIAL PUBLICATIONS DIVISION
ROBERT L. BREEDEN, *Editor*
DONALD J. CRUMP, *Associate Editor*
PHILIP B. SILCOTT, *Senior Editor*
RON FISHER, *Managing Editor*
WILLIAM R. GRAY, *Project Editor*
PENELOPE A. LOEFFLER, LOUISA MAGZANIAN,
    ELIZABETH L. PARKER, LUCY E. TOLAND, *Research*

Illustrations and Design
DON A. SPARKS, *Picture Editor*
URSULA PERRIN VOSSELER, *Art Director*
SUEZ B. KEHL, *Assistant Art Director*
MARGARET M. CARTER, CHRISTINE K. ECKSTROM,
    LOUIS DE LA HABA, JANE R. MCCAULEY,
    CAROLYN L. MICHAELS, JAMES H. MOONEY,
    Picture Legends
JOHN D. GARST, JR., PETER J. BALCH,
    CHARLES W. BERRY, LISA BIGANZOLI,
    MARGARET A. DEANE, JAIME QUINTERO,
    ALFRED L. ZEBARTH, *Art and Map Design,
    Research, and Production*

Production and Printing
ROBERT W. MESSER, *Production Manager*
GEORGE V. WHITE, *Assistant Production Manager*
RAJA D. MURSHED, JUNE L. GRAHAM,
    CHRISTINE A. ROBERTS, *Production Assistants*
DEBRA A. ANTONINI, BARBARA BRICKS, JANE H.
    BUXTON, ROSAMUND GARNER, SUZANNE J.
    JACOBSON, CLEO PETROFF, KATHERYN M.
    SLOCUM, SUZANNE VENINO, *Staff Assistants*
MARTHA K. HIGHTOWER, *Index*

Library of Congress CIP Data: page 199

*Billowing cloud marks the birth of Surtsey, a new volcanic
island that emerged from the ocean off Iceland in 1963. Lava
from a rift on the ocean floor pushed upward 425 feet to break
the surface. Page 1: Lightning crackles over Glacier National
Park, threatening a fire in the Apgar mountain range.
Endpapers: Cherubs of mythology create the earth's winds.
Bookbinding: An early weathercock faces into a breeze.*

OVERLEAF: SOLARFILMA. PAGE 1: B. RILEY MCCLELLAND. BOOKBINDING:
CHARLES W. BERRY. PAGES 4-5: N.G.S. PHOTOGRAPHER OTIS IMBODEN (UPPER);
HAWAII VOLCANOES NATIONAL PARK (MIDDLE); LEO AINSWORTH (LOWER)

*Nature's awesome powers:
Surging waves, roiled by
hurricane winds, assault the
Florida coast. In Hawaii,
visitors view an eruption
at Halemaumau Crater. A
tornado rips into Okla-
homa, trailing destruction.
Against such rampages of
nature, scientists marshal
research and technology,
attempting to predict them
and to lessen their
destructive effects.*

Contents

By Tom Melham

Earthquakes: Global Tremors, Drifting Continents

"And there shall be . . . earthquakes, in divers places."

—Matthew 24:7

TOM MELHAM, NATIONAL GEOGRAPHIC STAFF

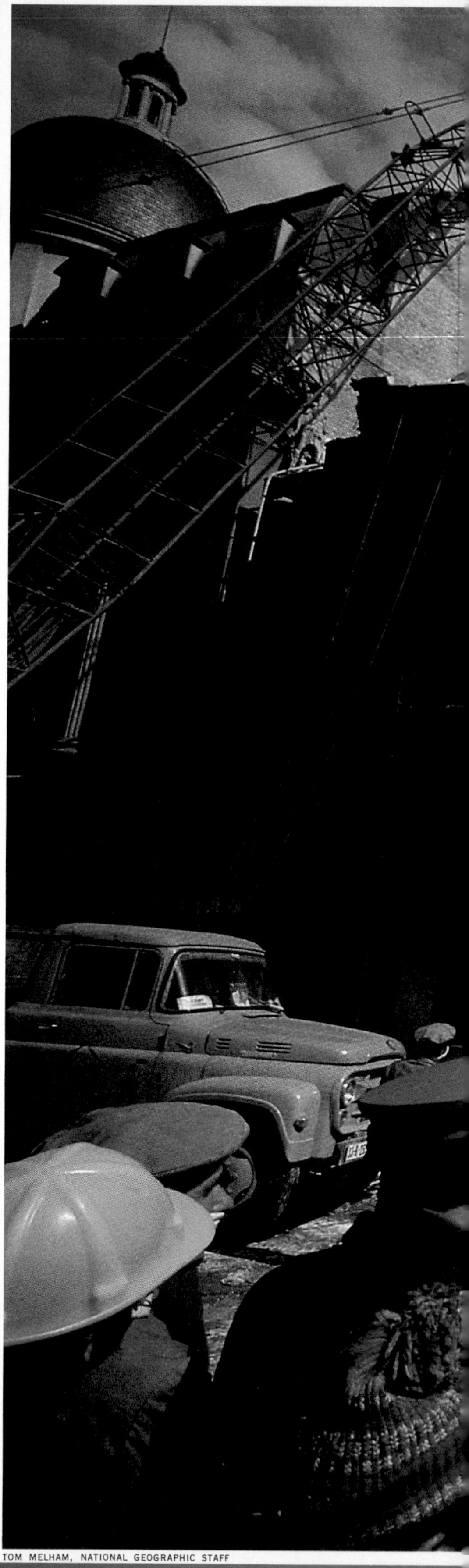

The shaking stopped in less than a minute. Retired schoolteacher Elena Enache lay in total darkness, conscious but stunned, gagging on plaster dust and wondering what had become of her normal, routine world. Only seconds earlier she had been knitting and watching television; then her high-rise apartment in Bucharest had shuddered violently, bouncing her up and down. Ceilings had given way; tons of debris had rained down—all just before the lights went out on March 4, 1977. Elena's eyes opened but saw nothing. She felt her bruised body and recalled the bouncing; her hands explored a wilderness of plaster shards and splintered furniture. Suddenly she realized the truth: An earthquake had leveled her apartment building—and she was buried alive.

Though in the heart of Bucharest, she was sealed off from it, trapped within a tiny, rubble-filled prison that lacked food, water, light, and even space enough to sit up. She realized that screaming was useless; no sounds could pass through to the outside world. Elena could see and hear nothing of the Bucharest she knew, nor of the present one convulsed with terror in the wake of the most destructive earthquake ever recorded in Romania.

In groping among the debris, she found a single familiar object—her transistor radio. As long as its batteries lasted, it remained Elena's talkative companion, sustaining her with reports of rescue operations. It also told her that tremors 30 miles to the north had crippled factories in oil-rich Ploieşti, setting local industry back several months. And 70 miles south of Bucharest, the Danube River port of Zimnicea lay in near-total ruins. Its mayor, Dumitru Sandu—who had taken office only minutes before the quake hit—suddenly found himself dealing with disaster instead of celebrating a political victory.

"The earth swayed round and round; I couldn't stand up," he told me when I visited Romania. "Eighty percent of my town vanished. We had no electricity, no gas, no water. Everywhere, I saw people crying. Some had to be pulled from the rubble; we didn't know if worse tremors would follow, so we took to the streets, where we lit fires and shared blankets to keep warm. Nobody slept that night for fear of aftershocks."

Fortunately, only five of Zimnicea's 20,000 residents died. The rest of the nation was not so lucky: Romania's earthquake of 1977 killed about 1,500 people, injured 10,500 more, and left tens of thousands homeless. Most of the victims were in Bucharest, where dozens of brick and concrete buildings crumbled. Many structures that remained standing bore great cracks and other scars; some stood sliced open like dolls' houses, spilling forth their contents onto the sidewalks; chimneys tilted crazily. For weeks the air hung heavy with plaster dust and chlorine—sprayed on the debris-strewn streets to prevent epidemic. Ambulance sirens merged with the rumbling of army half-tracks as soldiers, militia, and civilians worked around the clock searching for survivors, clearing wreckage, rerouting traffic, providing first aid and blankets. In many ways, Bucharest resembled a war-torn city.

I especially remember a stubby old woman in a long, black, peasant dress. Tears filled her eyes but refused to fall; they had become a permanent feature of her face. It was the sixth day after the quake, and she was standing at the edge of a heap of shattered masonry and twisted steel that had been the Continental apartment building. Rescuers clambered about the rubble in hopes of finding survivors trapped within. Already they had used "sniffer dogs"—trained in Switzerland to locate avalanche victims—but the dust had neutralized the dogs' sense of smell.

The old woman stood alone, as she had for days. From her worn purse she pulled a snapshot of a bride and groom. She stared at it a while, then quietly put it back. A few minutes later, she brought it out for another look. And then another. The smiling couple in the photograph— her daughter and son-in-law—had been inside the Continental with four other relatives when the quake struck. None had escaped; not even their bodies had been found. Yet each day and part of each night, the old woman returned with her photograph to watch and weep and hope.

She was not alone in her grief. Several blocks away, at another ruin, excavators dragged forth a lone shoe, some tattered rugs and stamp collections, a silver menorah—and three corpses, horribly mutilated and encrusted with mud. If anyone then had told me that a human being still lived within that jumble of brickwork, I would not have believed it. But somewhere in there, as yet unseen and unheard, a retired schoolteacher listened to her radio and patiently awaited release.

By now, Elena Enache had found that in addition to regular broadcasts she could pick up the walkie-talkie conversations of rescue teams. And so she knew her building was being picked apart brick by brick.

"You must survive," she told herself repeatedly. "Don't cry. Don't move. Don't sweat. You've got to save what water there is inside you. They'll find you—just wait."

It was in Elena's 188th hour of burial—nearly eight days after the quake—that she thought she saw a small circle of light in the darkness. Was she hallucinating? She could hardly move, but she managed to jab a loose stick at the luminous target.

Above her, rescue worker Miron Murea suddenly saw the stick's moving tip and raced toward it. He heard a faint request rise from the rubble below his feet: "Water."

Later that day, doctors would marvel at Elena's condition.

"She hasn't a single broken bone. She's in extremely good shape, the

Shattered mosaic and broken wall of a Romanian church reveal the power of the earthquake that struck the country on March 4, 1977. The mighty temblor damaged homes and factories across Romania— and shook tall buildings from Rome to Moscow.

Preceding pages: Five days after the quake, Romanians still work to clear the debris of a collapsed apartment house in downtown Bucharest. As firemen spray water on the rubble, a crowd watches— awaiting news of survivors.

9

Picking their way through rubble, Romanian soldiers carry an earthquake victim from the ruins of a Bucharest building (opposite). The quake of March 1977—the most destructive in the country's history—killed some 1,500 people and left tens of thousands homeless.

-SYGMA

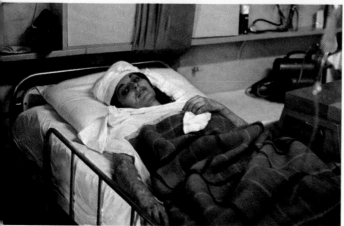

SIPA PRESS, BLACK STAR   TOM MELHAM, NATIONAL GEOGRAPHIC STAFF. SIPA PRESS, BLACK STAR (OPPOSITE)

Shaken but still standing, a 12-story apartment house sags above a Bucharest street corner. Crews worked around the clock salvaging belongings from condemned buildings (upper left) and searching for survivors. At left, Livia Negulescu rests in a Bucharest hospital. Trapped beneath wreckage with her fiancé—who died in her arms—she endured five days without food or water.

best of any earthquake victim I've seen," said Dr. Traian Bandila. Despite more than a week without food or water, his 58-year-old patient remained lucid.

Elena's round face was incandescent with smiles after her remarkable escape. She seemed to float above her hardship, and even said she could have held out for several days longer if necessary.

Other survivors were understandably demoralized. During the quake, several people had taken shelter in a doorway—and had watched helplessly as their apartment building, their possessions, their way of life collapsed about them. "It is like you are not even a child," one mused sadly. "You are nothing."

Indeed, the massive power of earthquakes has astounded mankind for centuries. The Old Testament holds that Sodom, Gomorrah, and other wicked "Cities of the Plain" vanished some 4,000 years ago in a fiery cataclysm—one that some scholars now believe was sparked by an earthquake which released vast amounts of natural gas and other flammables. San Francisco's famous 1906 quake caused a conflagration that wreaked far more damage than the earthquake itself.

But even without accompanying blazes, earthquakes are awesome. A trio of them destroyed Lisbon in 1755, killing a fourth of its 235,000 residents. The quakes of 1811 and 1812, centered at New Madrid, Missouri, were probably America's strongest ever, in terms of the area affected. Tremors were felt as far away as Massachusetts and South Carolina; in Kentucky, naturalist John James Audubon saw the land heave and ripple violently. The Mississippi River temporarily reversed its flow, creating Tennessee's Reelfoot Lake in the process. Thousands of aftershocks plagued the country for more than a year.

In 1897, a tremor hit India's easternmost state of Assam, destroying cities and towns over an area of nearly 8,000 square miles. The Guatemalan earthquake of 1976 left more than 23,000 dead and a million home-

less; in the same year, China's Tangshan quake near Peking claimed, incredibly, 655,000 lives. We still worry that tremors will disrupt the Alaska pipeline or deal San Francisco a repeat of the 1906 quake. In short, earthquakes remain terrifying and awesome. They strike unpredictably, with overwhelming strength and no visible cause, killing thousands of people one year and only a few the next.

Past civilizations often explained earthquakes through myths, most of which place the earth atop a giant animal or god that periodically twitches, causing its burden to shift. Algonquin Indians believed this creature to be a tortoise; the Japanese favored a catfish; Mongolian lamas thought it a frog. East Africa's Masawahili tribe concocted an ingenious balancing act that rivals the best of Ringling Brothers: A tremendous fish carries on its back a stone, which in turn supports a cow that balances the world first on one horn, then the other. Earthquakes result, of course, when the cow shifts the earth from one horn to the other.

Other legends blame the burrowings of a giant mole or the mischief of a subterranean demon. Aristotle thought earthquakes were caused by powerful underground winds. The Roman Emperor Justinian prohibited blasphemy and sexual misconduct on the grounds that they prompted earthquakes.

But what are earthquakes, anyway? A geologist will say they are brief tremblings of the earth's crust caused by a sudden movement or displacement of rock. Such movements release energy in the form of seismic waves (from the Greek *seismos*, "shock" or "earthquake"), which travel much like sound waves and generate vibrations—earthquakes—within the earth.

Anything that strains bedrock—volcanic eruptions, nuclear blasts, rapid changes in the water table, even another earthquake—can trigger a jolt. But most earthquakes occur for the simple reason that our planet is not the dead slab of rock many once imagined it to be.

Miles below our feet, contends the recent but widely accepted theory of plate tectonics, earth moves ceaselessly. Semirigid slabs, called tectonic plates, float upon a hot, plastic layer in the earth's mantle. They are huge — thousands of miles wide and 40 to 80 miles thick. Yet they move constantly. Like people jammed in a subway, the plates crowd each other edge to edge; a push at one point can set off a complex series of pushes and pulls among neighboring plates.

Although the plates move slowly in human terms (only inches per year), they have been moving for hundreds of millions of years — with spectacular consequences. Earth's continents and ocean floors are mere passengers affixed to the plates that — like giant conveyor belts — slowly push and pull them this way and that. But these crustal travelers are brittle; as they collide, separate, twist, or rub past one another, bedrock bends and breaks — and earthquakes result.

So it is that quakes generally occur in specific zones — those where plates abut. Romania's 1977 disaster followed a similar tremor there in 1940; both evolved from the convergence of the African and Eurasian plates. This same inexorable collision also generates earthquakes in northern Italy, the eastern Mediterranean, and Iran. Similar jostlings explain the seismic activity of Japan, California, Alaska, and other places.

And yet, earthquakes are only one effect. The interaction of the plates also spawns volcanoes, molds mountain ranges, expands some oceans at the expense of others, generates mineral deposits, and forms ocean trenches. It reworks earth's entire surface, and its effect on the continents is different from its effect on the thinner skin of the ocean floor.

For example, when two continental plates draw apart, the land gradually rips — as the surface of Africa is doing along its Great Rift Valley. When the plates collide, they usually thicken or fold the continental crust. Land masses have nowhere to go but up. Thus are born many of the world's major mountain ranges.

The general rule for continent-bearing plates, therefore, is that collisions breed mountains; separations cause deep valleys or ocean basins.

Beneath the oceans — where earth's crust is much thinner and denser — the opposite seems to hold true. As plates clash head-on, one subducts — dives under — the other and becomes absorbed in the mantle

"Nature is indifferent to the survival of the human species, including Americans."

—ADLAI STEVENSON

below. This creates deep ocean trenches such as those rimming the Pacific. And when plates separate, the thin seafloor crust rends and spreads apart, enabling molten material below to well up and form new crust that accumulates to become undersea mountains. One result is the Mid-Atlantic Ridge, a 12,000-mile-long mountain chain bisecting the floor of the constantly widening Atlantic Ocean.

Similar ridges, or "spreading centers," underlie all oceans, producing new crust while older material slowly disappears down the trenches. Thus the seafloor continuously recycles itself. But continental rock does not; because it is lighter, it remains at the surface and accumulates — making the continents much older and thicker than the ocean floors.

Where an oceanic plate rams into a continental one — as is happening today along South America's west coast — the ocean plate subducts. Land rises and ocean floor descends — leaving the side-by-side phenomena of towering Andes and 4,400-foot-deep Peru-Chile Trench.

In addition to colliding head-on, plates often meet obliquely, grinding against each other like millstones and causing visible cracks in the surface. Such is the case along California's San Andreas Fault, where the northward-moving Pacific plate rubs past the North American plate at one *(Continued on page 18)*

The Changing Face of Earth . . .

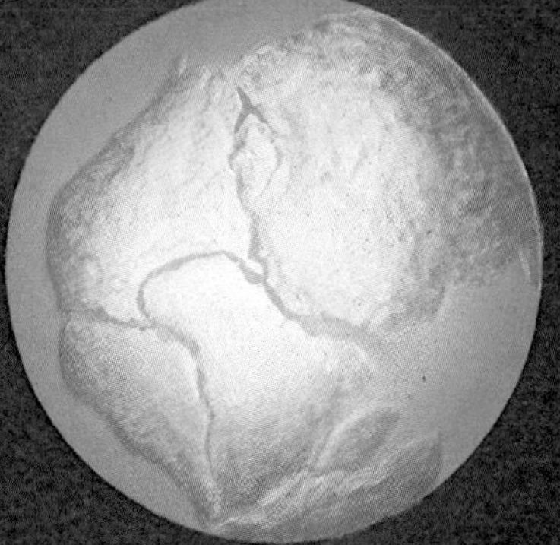

200 million years ago

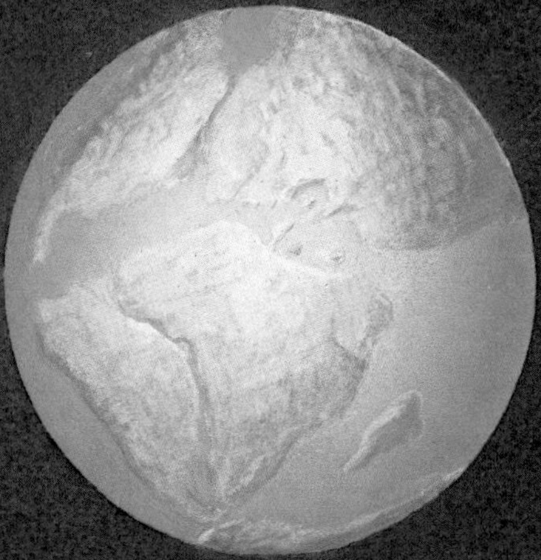

135 million years ago

65 million years ago

Today

Wheeling and drifting, colliding and cleaving, the huge landmasses we call our continents travel ponderously across the face of the planet. Throughout earth's long and turbulent history the continents have moved—bumping together to uplift mountains, breaking apart to create new shores. Even now they move, and the titanic forces shaping our lands will continue to reshape them for millions of years to come.

This new interpretation of our world—the theory of continental drift, or plate tectonics—explains the origins of our oceans and varied landforms, and reveals the sources of the powerful tremors and volcanic eruptions that wrack the earth.

From a reconstruction of the ancient supercontinent geologists call Pangaea, to predictions of what the planet will look like 50 million years from now, this series of paintings traces the majestic progression of the continents.

Two hundred million years ago, the world's landmasses formed one mighty continent, Pangaea—"all lands." German meteorologist Alfred Wegener, an early proponent of the continental-drift theory, coined the name. Waters of a universal ocean,

Panthalassa, wash Pangaea's shores; a remnant of that ancient sea survives today as our Pacific.

As millions of years passed, Pangaea slowly broke into continents. Some 135 million years ago, Pangaea had split into two great continents, northern Laurasia and southern Gondwanaland—separated by an embryonic Mediterranean Sea. A vertical rift jogs through Gondwanaland, marking the future coastlines of Africa and South America, and an Antarctica–Australia landmass breaks from its southern tip. India, severed from the east coast of Gondwanaland, moves north toward Laurasia.

Sixty-five million years ago, the Atlantic and Indian Oceans had formed. The Antarctica-Australia landmass edges toward the South Pole. Africa and South America drift farther apart as North America and Europe begin to rip along the future coast of Greenland. India eases toward Asia.

Today the Atlantic Ocean yawns wider by one or two inches a year. North America and Europe draw farther apart; Australia, broken free from Antarctica, heads north. The Americas have bumped together, and Arabia has eased away from Africa, widening the Red Sea. India, ramming into Asia, thrusts the Himalayas ever higher.

If the present rates of continental drift continue, scientists foresee more changes in earth's dynamic landscape. At right: two theoretical views of the earth's hemispheres 50 million years from now, with today's continents superimposed. East Africa has broken off along the Great Rift Valley, and the rest of the continent slides north, squeezing the Mediterranean. Rising sea levels flood Central America; the Atlantic and Pacific merge in the Caribbean, and the eastern shores of the Americas lie underwater. A sliver of California west of the San Andreas Fault, long detached from the mainland, slips north toward the Aleutian Trench. An inland sea fills central Greenland. Australia, not visible here, glides past Singapore, overrunning most of Indonesia.

Scientists can only postulate such awesome events, but they do know that our earth will change dramatically.

PAINTING BY JAIME QUINTERO

**Western hemisphere
50 million years from now**

**Eastern hemisphere
50 million years from now**

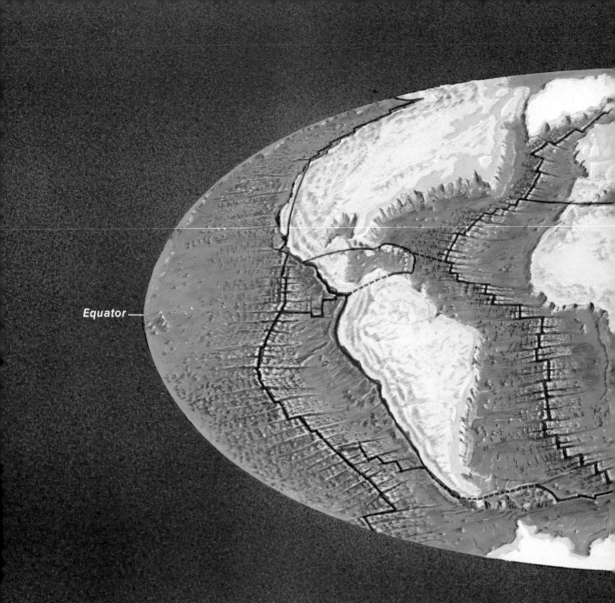

Equator

South America

Equator

... And Its Shifting, Fragmented Crust

Cracked into a global jigsaw puzzle of segments known as plates, earth's thin outer shell consists of our continents and ocean basins. Like shallow rafts afloat on a soupy sea, these huge crustal slabs drift over a plastic, semi-molten layer below, driven by steady churnings deep inside the earth.

Along their margins, plates jostle and grate, straining bedrock to the breaking point — and creating volcanoes and the tremblings we know as earthquakes. Charting frequent quake locations has helped scientists map the plates: Some consist of continents, or parts of them; others underlie the oceans. In a view of earth's fractured rind

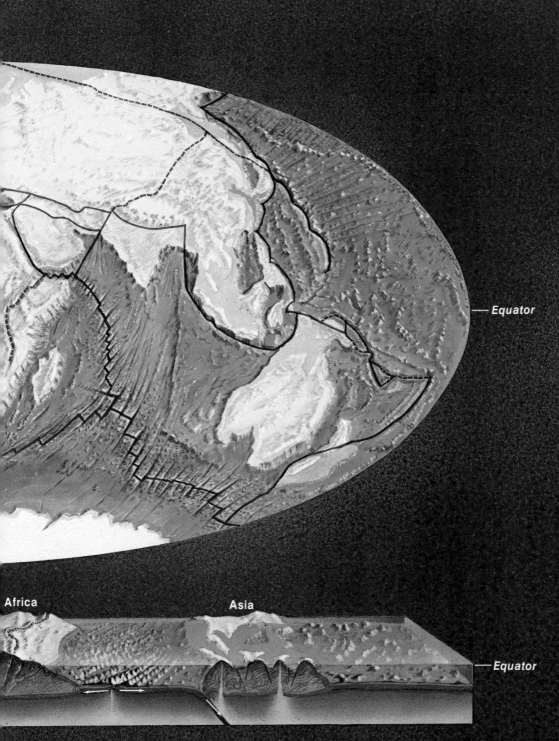

Equator

Africa

Asia

Equator

(top), with the oceans emptied, solid red lines trace the edges of plates; dashed lines mark uncertain boundaries. Along these margins mountains rise, volcanoes erupt, and islands burst from the sea.

A cross section sliced from along earth's Equator (above) exposes major plate boundaries and the activity associated with them. On South America's west coast, oceanic and continental plates collide. Subduc- tion pulls the heavier oceanic plate deep into the mantle, nudging the Andes ever higher — and fueling the volcanoes that erupt from their peaks. East of South America, the Mid-Atlantic Rift severs the ocean floor, widening the Atlantic and forming dikes of basalt that build new oceanic crust. The Great Rift Valley slices through Africa. In Indonesia, more subduction takes place, spawning the volcanoes of Sumatra and Java.

PAINTING BY JAIME QUINTERO

or two inches a year. Crustal rocks at first resist; friction and pressure mount continually. The rocks – being somewhat elastic – bend, compress, and stretch, but finally must give way. Seismic waves rush out from the point of sudden movement – and another earthquake hits California.

Slowly, the state's southwestern portion is sliding to the northwest. Along with it, Baja California has been drawing away from Mexico over the past 4 or 5 million years.

Such crustal movements are nothing new to the $4\frac{1}{2}$-billion-year-old earth. North America and the landmasses that eventually became Europe and Africa have repeatedly collided. One such collision 400 million years ago thrust up the Northwest Highlands of Scotland. About 270 million years ago, collisions produced a major mountain range, part of the ancestral Appalachians. And 70 million years after that – some 200 million years ago – the world's large landmasses joined into a single supercontinent known as Pangaea.

Since then, Australia and Antarctica have broken off and gone their separate ways. North America split first from Africa, then from Europe, creating the still-spreading Atlantic Ocean. Even today its shores parallel each other like the banks of a giant river – Labrador's rocky prow stands opposite Europe's Bay of Biscay; America's coastal profile from Maine to Georgia follows Africa's bulge; the close fit between South America and Africa has fascinated schoolchildren and scientists since the days of Sir Francis Bacon.

The movement of the plates explains other mysteries as well – such as why some now-distant continents share similar rock formations, coal deposits, and life forms; why in prehistoric times tropical ferns flourished in Antarctica and Greenland while glaciers covered the Sahara. These lands simply migrated from one climate zone to another.

The theory of plate tectonics, however, does not have all the answers. No one knows the details of the vast currents of heat that prompt the plates to move; we cannot yet predict or control earthquakes with any accuracy; despite all of seismology's recent advances, numerous misconceptions persist.

One is the popular notion that earthquakes cause the ground to draw apart and then snap shut like elevator doors, swallowing whatever – or whoever – happens to be above. It doesn't happen. A much more genuine danger occurs in steep, unstable areas of the world, where even minor quakes can trigger avalanches or mudflows, such as the one in 1970 that buried the town of Yungay, Peru, killing more than 18,000 people.

By their very nature, earthquakes are prone to exaggerated accounts. Eyewitnesses often see the same thing very differently. This is part of the trouble with scales of "earthquake intensity," which categorize quakes by the damage caused.

In 1902 Italian seismologist Giuseppe Mercalli developed a scale of earthquake intensity based on the observations of people in the quake area. Though since modified, his scale is still used for assessing earthquake damage at the site.

About forty years ago, Dr. Charles Richter devised a system that rated quakes independently of where they occurred and who observed them. "Seven-point-two on the Richter scale," said Walter Cronkite, as he reported Romania's 1977 quake. But what did it mean?

While not a precise measure of earthquake energy, Richter's scale permits a rough comparison of quakes. The scale is open-ended; very small quakes register minus numbers, and no maximum exists. But as yet no tremors have scored above 9.

Richter's scale is logarithmic, meaning that each unit jump in magnitude indicates a tenfold increase in seismic wave amplitude and roughly a 30-fold leap in energy. Thus a magnitude-8 quake releases about 30 times as much energy as a magnitude-7, about 900 times as much as a magnitude-6, about 27,000 times as much as a magnitude-5, and so on. These figures are only crude approximations, not strict mathematical formulas, emphasizes Richter—now a white-haired professor emeritus at the California Institute of Technology. Talking with me in his office, he added wryly:

"Actually, I'm surprised that the scale works as well as it does. It's based on an assumption that can't possibly be true—that you can compare earthquakes merely by multiplying some arithmetical factor. Every quake is unique; two magnitude-6.5 tremors can release appreciably different amounts of energy."

Originally devised for California quakes, the current Richter scale is a modified version suited to earthquakes throughout the world. Its 78-year-old innovator—a state resident since 1909—lives today near Pasadena, a few miles from the infamous San Andreas Fault.

"I'm not exactly scared out of my wits by the fault," he says. "The risk of earthquakes in California is greatly exaggerated. If people fear earthquakes, they shouldn't live here."

He has a point. Millions flock to California's scenic wonders, yet fail to realize that they are gifts of continental drift. Without tectonic plates abutting here, the Sierra Nevada

---

"Diseased nature oftentimes . . .
Is with a kind of colic
*   pinch'd and vex'd . . .*
Shakes the old beldam earth
*   and topples down*
Steeples and moss-grown
*   towers."*

—SHAKESPEARE, *I HENRY IV*

---

would not exist. Nor would one of America's greatest agricultural areas, the San Joaquin Valley. Nor would Napa Valley and its wineries, nor the Coast Ranges, nor San Francisco's famed hills. So it is with most dramatic landscapes: Alaska and Japan, the Himalayas and the Andes all are highly seismic, and owe their breathtaking profiles to the forces of nature constantly reshaping them.

If you're sold on southern California, advises Richter, don't try to escape earthquakes by avoiding the San Andreas Fault.

"Not only are there many other faults throughout California," he explains—indeed, they maze the southern half of the state as completely as do its freeways—"but risk isn't confined to the fault zone. Damage actually can be greater farther away, especially if the surface is loose sediment rather than hard rock."

Throughout the world, Richter adds, earthquakes have not increased in frequency, despite newspaper accounts to the contrary. True, they killed nearly 700,000 people in 1976. But most of those deaths followed a single jolt, the one in Tangshan, China—a densely populated city lacking earthquake-resistant buildings. The much stronger Alaskan earthquake of 1964 caused only 131 deaths—because it struck primarily wilderness.

"We usually have an average of one or two magnitude-8 quakes, worldwide, per year. Decidedly more major earthquakes occurred from 1896 to 1906 than in any decade since. There hasn't been a serious quake on the San Andreas Fault since 1906, and the part of the fault nearest Los Angeles has been nearly dormant since 1857. Today, we're probably overdue."

The San Andreas Fault stretches about 700 miles from Cape Mendocino to the Salton Sea and is "probably several tens of millions of years old—maybe older," Caltech geologist Clarence Allen told me. Though not the world's longest fault (China's Altyn Tagh system exceeds it), the San Andreas is the best instrumented and most studied in the world.

"The great bulk of this country's seismologists and earthquake engineers," Allen notes, "live near San Francisco or Los Angeles—two areas most prone to earthquakes." The seismologists monitor opposite sides of the San Andreas Fault for changes in tilt, strain, displacement, and other warning signs. Some are studying a 32,400-square-mile bulge northeast

(Continued on page 24)

MAP BY PETER J. BALCH AND MARGARET A. DEANE

Principal disturbance area
Minor disturbance area
● Recorded point of vibrations and sounds
● Recorded point of vibrations only

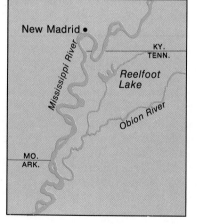

New Madrid: America's "Forgotten Earthquake"

Severe but little-known quake shook the frontier settlement of New Madrid, Missouri, on December 16, 1811. From Indian camps on the Kansas plains to the streets of Boston, a million square miles of America's land shivered with tremors. In Kentucky, naturalist John James Audubon wrote that "the ground waved like a field of corn before the breeze." One steamboat captain saw the mighty Mississippi briefly reverse its flow as its waters heaved and fell with the shaking. Tiny New Madrid probably stood at the epicenter—the point on earth's surface directly above the source of the quake. In northwest Tennessee, 20 square miles of swampy woodlands sank; water seeped into the soggy depression, creating Reelfoot Lake (right).

NATIONAL GEOGRAPHIC PHOTOGRAPHER VICTOR R. BOSWELL, JR.

21

of Los Angeles that has uplifted as much as a foot and a half during the past 15 years. Others focus on horizontal motion; at Hollister—about 100 miles southeast of San Francisco—one side of the fault grinds past the other at slightly more than an inch a year, offsetting fences and splitting buildings that extend across it. Such regular movement, accompanied by small earthquakes, is called "creep."

But surface rocks near San Francisco and Los Angeles show no displacement, although the plates probably move at the same rate as at Hollister. Curiously, the fault plane "locks up"—pressures build over the decades until a major quake results. Why doesn't it "creep"? Allen observes that the fairly straight San Andreas Fault changes course near both cities. The curves may cause the system to balk. "Rock types also could play a role; areas where we see creep often have a lot of serpentine and other soft, slippery rocks—while areas in both Los Angeles and San Francisco adjacent to the fault are underlain by granite, which is relatively brittle."

Not all the world's earthquakes are nature's doing. Nuclear tests in Nevada have triggered thousands of small quakes. During the decade after completion of Hoover Dam and Lake Mead, some 6,000 minor tremors shook a region that had no prior history of quakes—presumably because the addition of the weight of Lake Mead's 255 square miles of water exerted new strains upon earth's crust, or because the water "lubricated" faults beneath the lake. Between 1962 and 1965 a flurry of quakes near Denver followed the injection of liquid wastes into a well-hole. The pumping was discontinued, and the quakes eventually stopped.

Such liquid injections apparently lubricate the fault planes, freeing locked areas somewhat. The result—more earthquakes but usually small ones—holds some tantalizing possibilities, such as earthquake control. What if deep holes were drilled near San Francisco and then periodically filled and emptied: Would this enable the locked bedrock to move in many small steps rather than in one big one, thus eliminating earthquakes of 1906 proportions?

Well, it's possible—in theory. By enabling the San Andreas Fault to slip an inch or so about every four miles just once a year, the average rate of fault displacement would be accommodated. People living near the fault would feel perhaps ten shakes a year strong enough to get their attention—but not strong enough to do any damage.

Geologist Allen suggests an alternative he considers more achievable: "If, instead of trying to eliminate the big ones, we could control them enough to make them occur on schedule, we'd be able to evacuate people from hazardous structures beforehand, empty endangered reservoirs, and take other precautions that would minimize losses. This is definitely worth dreaming about. Not totally absurd, anyway, but it's a long, long way off. There would be incredible legal and economic problems: When an earthquake is no longer an act of God but of man, who's responsible for the damages?"

As earthquake research proceeds, California continues to await the inevitable crusher. Millions of residents simply ignore it. Others capitalize on it: San Jose named its professional soccer team the Quakes; Swensen's Ice Cream Factory—based in San Francisco—offers a gargantuan $5.50 sundae known as "The Earthquake": eight scoops (about two pounds of homemade ice cream),

Preceding pages: A jagged rift splits Iceland, emergent plateau of an undersea mountain range. Along the fissure, two crustal plates slowly part—and Iceland grows, widening nearly an inch a year.

Where crustal plates collide: The San Andreas Fault (infrared photograph) runs through much of California. As the western plate, on the right, grinds north, earthquakes rock the land.

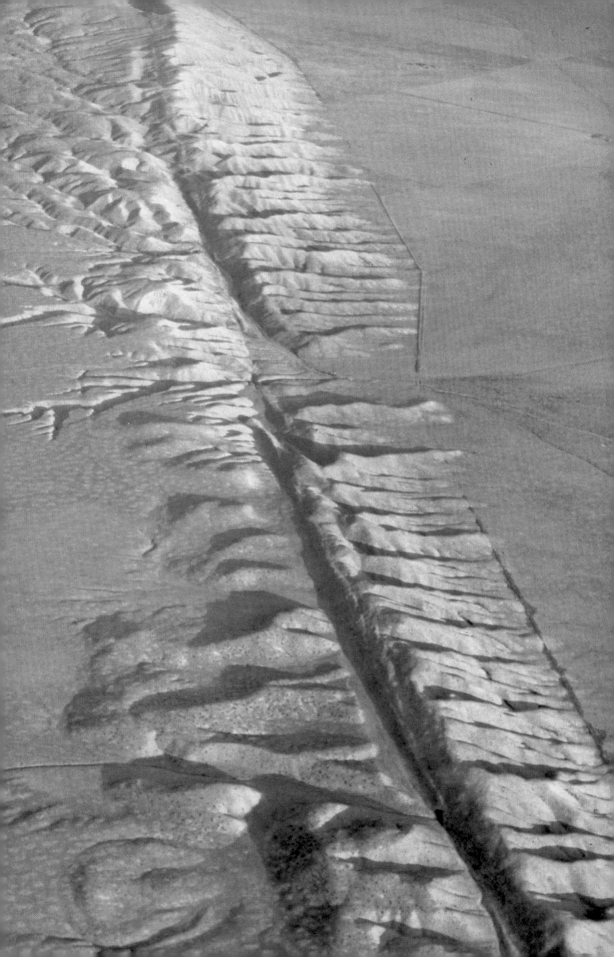

*Ruined by earthquake and fire,
San Francisco smolders in the
wake of the violent tremors of April
18, 1906. One woman awoke to the
sounds of "chandeliers shattering
through the house." The massive
shock probably would have
reached 8.3 on the Richter scale
— a system for rating earthquake
strength. At right, Charles
Richter, developer of the scale,
studies seismic readings in his
California office. Today San
Francisco enforces strict
building codes, and residents hope
that structures like the towering
Transamerica Pyramid (left) will
be "earthquake-resistant."*

with eight toppings plus almonds, whipped cream, and cherries. Promises the menu, "Your taste buds will register 8.9 on the Richter scale."

San Francisco cartoonist Dan O'Neill keeps in touch with California's number one natural hazard by inviting fellow citizens to join the "Free the San Francisco Earthquake Committee" he founded in 1969.

"Every now and then," he explains, "some occult person out here 'predicts' an earthquake, and a lot of people get scared. I felt I had to do something; that's how the committee got started."

Its members do not meet regularly, but are instructed to "free" the earthquake by jumping off chairs at exactly 8 a.m. every Tuesday. Each landing sends a very minor shock into the earth, O'Neill says, with tongue in cheek, and if the membership ever swells into the thousands, the combined wallop someday may be large enough to trigger small quakes, thus avoiding a bigger one.

While O'Neill's friends may disagree over his seriousness of purpose, none disputes that his committee is part of California's booming earthquake culture—one that includes not only ice cream and soccer but also thriller movies, manuals on tremor-proof home construction, and psychics seeking to link mind control with quake control.

Far different earthquake cultures thrive on the Pacific's opposite shore, where a combination of dense populations and high seismic activity makes for especially deadly quakes. China's earthquake prediction program relies on a vast, well-organized network of amateurs—farmers, schoolchildren, housewives, and others—who watch for changes in the earth and in animal behavior, and report them to seismologists.

Many Tokyo department stores market various models of *bōsai-bukuro*—"emergency preparedness kits" designed to help earthquake victims survive the chaotic aftermath. The more extravagant kits contain all the comforts of a 1950's fallout shelter—food and water, emergency power and lights, stoves, blankets, radios, rope ladders, hard hats, portable toilets, and more—indicating just how seriously the Japanese take their earthquakes.

They should. Japan, though slightly smaller in area than Montana, supports 111 million people—nearly 150 times the population of that state. Its earthquakes similarly outnumber Montana's, for along Japan's eastern shore an edge of the Pacific plate is moving under the Eurasian continent. One result is the five-mile-deep Japan Trench. Another is that the nation is compressed by the powerful forces a few inches each year, then expanded again, generating strains that make the Land of the Rising Sun also a land of frequent earthquakes.

Even its small ones can be debilitating, in unexpected ways. The mountain-rimmed town of Matsushiro, about 120 miles northwest of Tokyo, has weathered as many as 10,000 quakes in a single day.

"Most were too slight for humans to sense, but some days we felt six or seven hundred—one every two minutes or so," seismologist Dr. Tsuneji Rikitake told me. "At first, people couldn't sleep or do anything. The ground was continually vibrating."

In vain, Matsushiro's inhabitants tried to adapt to what would become the world's longest recorded earthquake "swarm"—a sustained period of many small quakes—that began in August of 1965. Veterans of previous swarms predicted that this one soon would quiet down.

It didn't. The quakes continued to multiply. Occasional magnitude-5 or 5.5 shocks further roused the community. Incredibly, the earthquake count reached 450,000 within the first year! Were they only foreshocks preceding a much larger quake? No one could say.

Stoves and furnaces were shut off to minimize fire damage in the event of a major quake; farmers slept in their fields to avoid the danger of collapsing homes; new mineral springs put forth, and existing hot springs turned hotter. Reports of ominous earthquake-related sounds and lights increased. Schools taught evacuation procedures. Eventually the town's trial-by-shaking began to crack building facades and the determination of some residents. A mobile clinic was set up to treat the growing number of insomniacs and tension-sufferers. Matsushiro neared financial and psychological crisis.

Only after a year and a half did the swarm begin to subside. Even today, 12 years after its onset, seismic activity remains "abnormally high," notes Dr. Mamoru Katsumata, chief researcher at the Matsushiro Seismological Observatory.

"From the start of the swarm until yesterday," he told me in June of 1977, "we've counted 720,665 earthquakes. But only 63,104 were large enough for humans to detect. And even then, the combined energy of all quakes since 1965 totals less than that of a single magnitude-6.5 tremor." Today, Matsushiro's inhabitants feel merely 30 or so quakes each year. Still, that's well above the pre-swarm rate of two or three annually.

"So, seismically, we still have a swarm. But in the social sense, it's over. No one thinks anything of the occasional quakes we have now."

In fact, seismic areas with *fewer* earthquakes than normal often cause greater concern among scientists. Such "seismic gaps" may indicate spots where pressures generated by the plates are not being relieved. From Tokyo south for some 180 miles, major quakes have shaken Japan's east coast about once every 85

DAN O'NEILL

Whimsical answer to the earthquake threat: A cartoon strip laughs off the widespread but illogical notion that San Andreas quakes may send California sliding into the Pacific.

Surprise shocks wrench the suburbs of Los Angeles: On February 9, 1971, the earth suddenly moved along a minor fault in the San Fernando Valley area north of Los Angeles. Measuring a "moderate" 6.5 on the Richter scale, the earthquake released more energy than the Hiroshima atomic bomb. At the Sylmar Veterans Hospital (below), two old buildings collapsed like houses of cards; as night falls, rescue teams search for survivors in the crumbled structures, where 45 people lost their lives. Stairwell wings lie toppled (far left) at the new earthquake-resistant Olive View Hospital. At lower left, a wheelchair dangles from one of the fallen wings. The tremors shook goods from supermarket shelves (left), buckled freeways, and weakened dams, causing half a billion dollars' damage in suburban Los Angeles. Though emergency efforts proved efficient, the tragedies focused public attention on the need for sturdier buildings and better planning.

Whiplashed by shock waves, an Anchorage street attests the force of North America's strongest recorded earthquake. Measuring 8.5 on the Richter scale, the quake rocked southern Alaska in 1964. Below, fissures slice an Anchorage school —empty when the tremors struck.

years. But the region southwest of Tokyo has not had a major tremor since 1854—a lapse of 123 years.

"There's a very real danger here," warns Rikitake. "We have every reason to expect a big earthquake, one as devastating as a war. There's a nuclear power station on the coast, and many high-speed trains. Just imagine what might happen if a quake parted the rails under a train traveling 125 miles an hour, or if it cracked that reactor. A major shock may not destroy modern 'earthquake-resistant' buildings, but it could sway the top floors as much as six feet, throwing desks into people, breaking pipes,

and making the buildings unusable.

"And what of earthquake-caused fires? There's nothing we can do to prevent broken water and gas lines. We can only ask the people to put out any fires they see."

"Our primary problem is fires, not the quakes themselves," agrees Ichiro Koshii of the Earthquake Preparedness Division in Shizuoka. "So we're urging citizens to install devices that would automatically turn off their kerosene stoves whenever the earth vibrates."

Present school programs, he adds, teach children how to react to earthquakes: (If you're inside, seek

shelter under a desk. If you're outside, stay in open spaces away from buildings. Don't drive; you'll only cause traffic jams and block fire engines. Above all, don't panic.) Television broadcasts instruct viewers to set aside caches of water and food. Shizuoka also educates its people with a mobile earthquake simulator.

At first glance, it looks like a misplaced camper jammed atop a fire truck. The truck — painted bright red because the simulator is maintained by the city fire department — supports a square room complete with doors, windows, and a foot-wide "front yard" of artificial turf. Various interior props lend it the decor of schoolroom, home, or office — the idea being to teach people how to respond to earthquakes in familiar surroundings. The day I arrived for my simulator "lesson" must have followed a session with housewives; the room contained a kitchen table and hanging lamp, a two-burner stove, some shelves, and a beaded curtain.

The simulator mimics earthquakes of levels 4, 5, and 6 on the Japanese intensity scale of *shindo,* which rates quakes according to the vibrations actually felt. Intensity-6 includes the world's worst recorded quakes, ones hefty enough to split the

Huddled in prayer before the ruins of their
church, survivors in the Guatemalan town of
San Martín Jilotepeque await food and
medical aid following the devastating
earthquake of February 4, 1976. Landslides
triggered by the quake sheared mountain
faces (below, far left) and clogged roads to
stricken villages. Highways cracked, railroad
tracks buckled, and one major bridge caved
in — the Agua Caliente (below, center), on the
supply route to the Atlantic. Relief poured in
from the U. S. and nearby countries; above,
rescue workers prepare tamale batter at a field
kitchen set up by the Mexican government.
Outside Guatemala City, Red Cross tents
shelter the homeless (below). Twenty-three
thousand Guatemalans died.

J. P. LAFFONT, SYGMA (BELOW). NATIONAL GEOGRAPHIC PHOTOGRAPHER ROBERT W. MADDEN

earth, spawn landslides, and level cities. Having never experienced a quake of *any* size, I wondered silently as I entered the simulator: *(Do I really want to do this?)*

I immediately felt the slight vibrations of its motor, a sensation similar to most minor earthquakes, I was told. *(Nothing to it, thought the novice. Bring on the real thing.)*

As the attendant shifted gears to intensity-4, my "kitchen" began a slight back-and-forth motion. *(No problem.)* The shaking increased at intensity-5, moving the room not only horizontally but also up and down. Low, growling "earthquake sounds" came to my ears. The beaded curtain swayed. *(No worse than Coney Island—I'm still on my feet.)*

The shift to intensity-6 brought quantum leaps in both vibrations and noise. Growls became jackhammer-level roars; thrown off balance, I grabbed for the table and watched it skid across the rumbling floor. The hanging lamp overhead swung like a berserk Tarzan. Strands of beaded curtain whipped at my head as I dodged; a stove burner flew off while the stove itself crept away from the wall and headed toward me. *(This is insane! You mean you people put up with this madness year after year?)*

It was all over in a few seconds, just like the real thing. Koshii explained afterward, "Your first priority is to protect yourself." Fine, I thought, recalling that the "earthquake" had turned my thoughts totally inward. All I'd wanted to do was to stay unhurt and on my feet.

"But the second priority is to turn off the stove and grab a fire extinguisher." I hadn't. In a real quake, this mistake could have sparked a holocaust. Next time, I decided, would be different.

Fate wasted no time in testing me. The very next morning, as I lay in bed in Tokyo, a real earthquake struck. It was only intensity-4, but I was already awake and knew immediately what was happening. The bed's back-and-forth sways perfectly duplicated those of the simulator; drinking glasses on the table quivered and clinked; the ceiling made slight cracking noises, although the building was steel-reinforced.

My first earthquake lasted only a few seconds, and I'm ashamed to admit that, again, I flunked the test. I neither sought the strongest part of the room nor looked for a fire extinguisher. I just stayed where I was and waited for the ceiling to fall.

A glimpse from my window showed that Tokyo's traffic proceeded normally; in a nearby high rise I counted only three people curious enough to look out their windows. Visits to major department stores later that day showed that the quake had had absolutely no effect on lagging sales of earthquake kits. Though I waited at one display for half an hour, not a single person showed any interest.

Tokyo's last devastating earthquake—the magnitude-8.3 quake of 1923—ripped through the city just as thousands of stoves were being lit to prepare the noon meal. Many of the wooden buildings collapsed at once; fires—whipped by strong winds—swept through the city for nearly two days. Some 40,000 homeless crowded into a vacant 17-acre lot and found themselves trapped between Tokyo's Sumida River and the advancing wall of fire. About 38,000 drowned or were burned there; another 20,000 perished throughout the city; 41,000 more died in Yokohama and elsewhere.

Masayoshi Satow, then a sophomore at a university in Tokyo, was vacationing about 60 miles away on the day of the quake. He returned two weeks later—on foot because the quake had damaged a rail line—and looked across the city from Ueno Hill.

"There was nothing!" he told me.

Sea of mud and rock obliterates the town of Yungay below Peru's Mount Huascarán. On May 31, 1970, an earthquake snapped a wall of ice from the mountain face, setting off an avalanche that killed more than 18,000 people.

Tsunami: Surging Walls of Water

Racing across oceans at jet-plane speeds and building to mountains of water near shore, tsunami lash
earth's coastlines with awesome force. When an earthquake occurs near or under the sea, ocean waves radiate from the shock's source in widening circles. In the painting at left, the rupture of a vertical rift in the ocean floor generates waves that rush toward a coastal city. In the open sea, these waves barely ripple the surface — undetected by airplanes or ships. As they reach shallow waters, the waves slow down and pile up in crests sometimes more than 100 feet high. At upper left, churning waters break over a seawall on the north coast of Oahu during a 1957 tsunami. Undersea landslides, volcanic eruptions, and man-made explosions can also trigger tsunami. The scoured waterfront of Valdez, Alaska (above), reveals the power of a tsunami that hit during the 1964 Alaska earthquake.

"You could see all the way from Ueno to Ginza to the sea, six miles or so. The whole city was only this high." Satow pointed to a spot halfway down his calf. "Most of the buildings had been of wood; only ashes were left."

As if the horror of earthquakes isn't enough to bear, Japan also must cope with another equally terrifying force of nature — *tsunami,* or "seismic sea waves" — that can race across oceans at jet-plane speeds and crest in walls of water more than a hundred feet high. One in 1771 tore massive coral heads — some weighing as much as 750 tons — from reefs near Japan's Ryukyu Islands and flung them ashore, several miles from the ocean, where they remain today. The main wave, reported to be 260 feet high, killed 11,000 people in the Yaeyama Islands.

Though traditionally called "tidal waves" in the West, tsunami have nothing to do with the tides; they stem from earthquakes, volcanic eruptions, landslides, and man-made explosions near or under the sea. Like ripples set up when a pebble is tossed into a pond, they rush out in widening circles. The distance from one wave to the next averages a hundred miles, but the waves stand only a few feet high on the open sea, usually invisible to airplanes or ships. Unlike wind waves, their speed is determined by ocean depth; deepwater tsunami travel at more than 500 miles an hour.

In shallows, however, they rapidly slow down and pile up in awesome crests that rip brick homes off their foundations, clip dense forests into a stubble of stumps, cave in huge oil storage tanks, and heave ships miles inland. These crests alternate with equally powerful troughs, which cause the sea to recede from shore for hundreds of yards. Currents charge in and out during a tsunami attack; debris thrown inland by one crest may get sucked out to sea in the next trough.

The most lethal tsunami ever recorded followed the eruption of Krakatoa in 1883. In a series of immense explosions, this volcanic island between Java and Sumatra belched

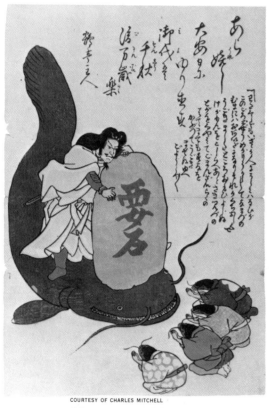

COURTESY OF CHARLES MITCHELL

Mysterious quakings deep inside the earth baffled ancient peoples, and many civilizations created myths to explain the rumblings. One Japanese legend attributes the tremors of earthquakes to the capricious floppings of a giant catfish living at the bottom of the earth. Fearful people pray to an "earthquake god" who tries to restrain the fish with a huge stone.

forth some five cubic miles of rock and debris; small particles were blown as high as 50 miles and orbited the earth for months. What was left of Krakatoa collapsed in upon itself. Tsunami ranged the Indo-Pacific at reported heights of up to 120 feet. Whole islands were wiped clean of life; 36,000 people perished in Java and Sumatra alone.

Some historians have suggested that a seismic sea wave in the Red Sea thousands of years ago may account for the Biblical story of Exodus, in which the Red Sea opens up (water receding before tsunami?), then abruptly closes (the tsunami?), thus enabling Moses and his Israelites to pass out of Egypt, and drowning Pharaoh's pursuing army.

Japan's 1896 tsunami claimed more than 27,000 lives and washed away some 10,000 homes and 7,000 fishing boats. Little wonder that U. S. Navy Captain W. R. Wickland "got an eerie feeling, like walking into a morgue" the day he saw a tsunami approach Hawaii in 1946.

"It was right after the war," he told me, "and I was the harbormaster, pilot, and district manager in Hilo. I was aboard the *William D. Hoxie,* a Liberty Ship, moored at Pier Three. At about ten to seven in the morning, I noticed the ship drop a bit. All our lines went slack. Then she rose. It happened again, then a third time — and all the lines parted. I looked out and saw what looked like a low, long swell at sea; way out, but coming in awfully fast. Seemed like three separate waves, each behind the other, came together in one monster wave. I was on the *Hoxie's* upper bridge, some 46 feet above waterline. That wave was just about eye-level and probably two miles long. I never expected to see anything like it again.

"We tried to back out from our slip, but before we got very far the whole harbor emptied — that big wave was coming in and there we were, lying on the bottom! It's really strange to look out and see coral reefs around a ship — reefs that are normally underwater. That first wave took about 7,500 feet of a 10,000-foot-long breakwater with it. The cap

rocks were 12 tons apiece; when you strip something like that, mister, you've got *some* power. I watched it slam into Pier One nearby and heard a great crunching sound as the eight-by-eights and twelve-by-twelves broke and went inland, as if an enormous hand had just crushed the pier and swept it away.

"Most tsunami aren't breaking waves; they're more like low swells. But that one broke — and tore up everything it touched. My office went; some Coast Guard boats flew by — one wound up inland of some big molasses tanks — and a yacht was thrown up to the main highway. Every structure, building, piece of equipment on shore seemed to take off.

"Terrific currents pulled at the *Hoxie,* changing direction all the time, doing as much damage going out as coming in — that's when I had the most trouble piloting her. But luckily, even though we were grounded, our stern was pointed out to sea, helping us ride out that first wave. Coral reefs were not far away on either end. We had 1,700 barrels of high-test gas on board and just wanted to stay off that reef — I was sure we'd hang up on it. Still don't know why we didn't. *Four times* that harbor filled and emptied before we got out."

Fourteen years later, a tsunami again roared into Hilo Bay. Again, Wickland was there.

"I guess that one was smaller than the one in '46, but to me it was even more impressive. It came at night. I was three miles from downtown and still heard the electricity blow, barking like a 90-millimeter gun. Sparks flew everywhere. I heard buildings crunch as if they were being chewed. When I got to see them, they looked like they'd been in a tornado — just a bunch of exploded matchsticks. The main wave didn't hit us straight on; it deflected off the cliffs. Water went 40 or 50 feet up the cliffs near the harbor entrance, then sloshed back over the town and nearly wiped it out."

The same tsunami — spawned by a magnitude-8.5 quake in Chile — also smacked Japan, bounced off, and headed back. "The ocean agitated for

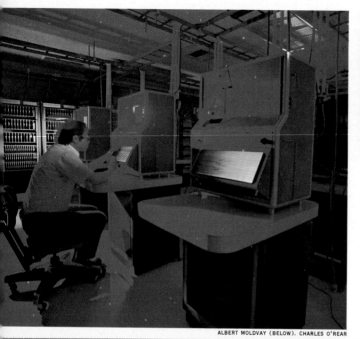

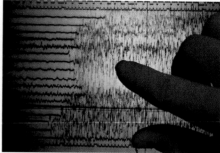

Seeking clues to future shocks, geologists study tremor patterns along the web of faults that laces California. At the U. S. Geological Survey laboratory in Menlo Park (left), a scientist analyzes signals from seismic stations across California; oscillations on a film record vibrations from 17 stations, indicating a small quake (above).

Firing laser beams across the San Andreas Fault (above), seismologists record each shift of rock; a plane checks temperature and humidity, which affect the beam. In a simulator (opposite), slabs form a "miniature fault." Dozens of tiny sensors show in detail how the rock deforms before and during quakes.

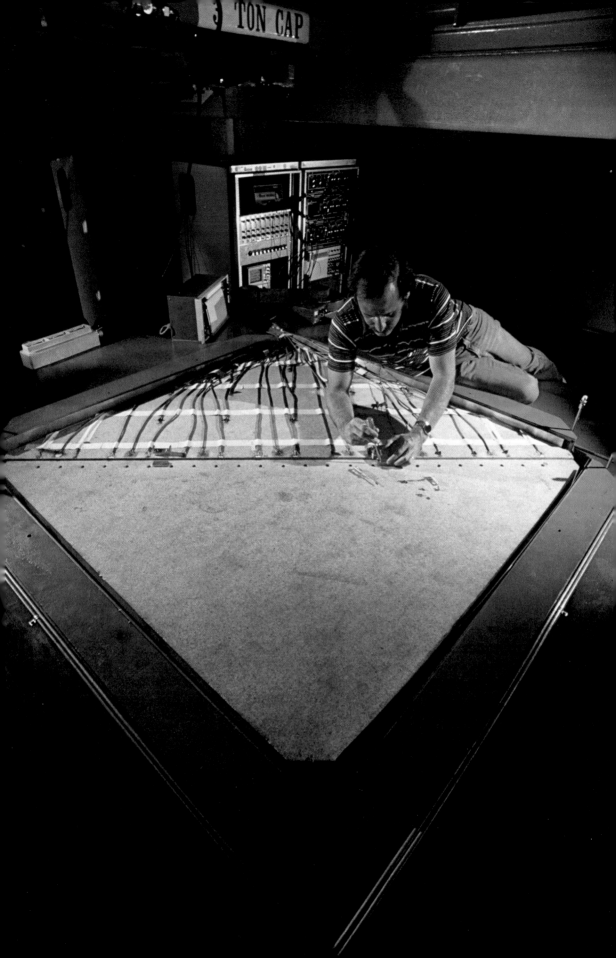

days," recalls Dr. Kinjiro Kajiura of Tokyo's Earthquake Research Institute. "It was South America's biggest recorded tsunami."

What made it especially tragic, however, was that most of the 138 deaths it caused in Japan could have been averted. Seismologists throughout the world detected Chile's 1960 quake within minutes — yet for some reason Japan's seaside communities were not evacuated. Even at 500 miles an hour, waves need about 22 hours to cross the Pacific — allowing plenty of time to issue warnings. But because previous South American tsunami rarely affected Japan, no one expected it to be so severe.

Today, the Pacific Tsunami Warning Center, located in Honolulu and administered by the U. S. National Weather Service, tries to minimize the loss of life and property from tsunami by monitoring seismometers and tide gauges throughout the Pacific. A major quake anywhere in this area rings warning bells; geophysicists soon determine the quake's location and size. If it is likely to generate tsunami, they issue a "watch" to all member nations in the Pacific. When observers or instruments detect actual tsunami waves, that "watch" becomes an official "warning," updated as the waves proceed. Dr. Eddie Bernard, former director of the Center, explains:

"It's like making a weather forecast — we know a wave is, say, one foot high on an island 3,000 miles away and want to know how big it'll be when it reaches other shorelines in the Pacific. But that's not scientifically possible. Sometimes, areas such as coves well in the lee of tsunami are hardest hit because of a wraparound effect." The waves can reinforce each other in even stranger ways. Sometimes, tsunami deflect off a shoreline into approaching waves, decreasing their force. But when the crests and troughs of these opposing waves coincide, wave heights — and potential destructiveness — actually increase. The third or fourth wave of a tsunami may be the largest because of wave dispersion and interaction.

With continued advances — satellites to speed communications and computers to analyze data — Bernard hopes eventually to attain a fail-safe system: no missed tsunami and no false alarms. Even the Center, however, cannot warn of locally generated tsunami that can strike within minutes of the earthquake. Such a wave ravaged the Philippines in 1976; Syd Wigen of the International Tsunami Information Center, which assists nations in organizing local warning and evacuation procedures, told me about its aftermath:

"Five thousand people died. You can talk about statistics and they don't mean a hoot, but it's different when you know the people. Every day I'd meet someone I knew who had lost relatives or friends, and I'd go cry my eyes out. It gave me an awareness of the human situation, an awareness that reading never would.

"But what really gets me is that most of those deaths were needless — the people just didn't know how to react. They don't understand tsunami; they don't know that when they feel an earthquake they should take to the hills as fast as possible. They're fishermen, living on the beach or in huts just above the water, totally exposed to the waves. Today, they've rebuilt their homes, all in the same areas. Those places got whacked fifty years ago, too. Looks like the next time will be just as bad. But it doesn't have to be that way."

One method of forecasting local tsunami is to predict the quakes that cause them; the only hitch is that earthquake prediction is in its infancy — and may never grow up.

Dr. Richter does not object to

Can animals sense an impending earthquake? Geologist Ruth Simon conducts tests with cockroaches in California; a sensor that detects their movements has recorded increased activity before quakes. Such observations of unusual animal behavior may one day serve as a prediction tool.

serious prediction research, but deplores the activities of "mystics, crackpots, and publicity hounds." He is not alone in considering earthquake *preparedness*—improved education and building procedures—more achievable than *prediction*. Yet other seismologists around the world invest their time—and a lot of money—in prediction research. Americans roam paper jungles of data, hunting for warning signs in changing water levels, land movements, animal behavior, magnetic and electrical fields, and earth's internal strains. Soviet scientists watch for upsurges in radon—a radioactive element found in groundwater—that seem to precede tremors. Japan uses all of these methods. So does China, which successfully predicted its 1974 earthquake in Haich'eng.

The Survey's Dr. Barry Raleigh, who headed a scientific delegation to China, attributes the Haich'eng triumph to force of numbers:

"At least 10,000 full-time workers (America has a tenth as many) and thousands more amateurs study the earthquake problem. There are no major differences in technology; they just have a lot more people—and a lot more earthquakes. Also, they have the advantage of a couple of thousand years of historical observations. Even the peasants take quakes seriously, and are very knowledgeable. When a local well suddenly went artesian, all the people just packed up and moved out, about a day before the Haich'eng quake, before any official evacuation was announced. They were only farmers, but they knew what to do—and they saved a lot of lives."

Dr. Jack Evernden, a geophysicist with the USGS, points out, however, that Haich'eng's highly publicized success followed a not-so-publicized false alarm at Pan Shan, which forced hundreds of people to flee their homes in the Manchurian winter. And Chinese seismologists failed to foresee the 1976 Tangshan disaster that killed two-thirds of a million people. Clearly, earthquake prediction has much room for progress.

One popular prediction tactic concerns the supposed ability of animals to sense impending tremors. Tales both past and present abound.

"It's all anecdotal material," says Evernden. "People say, 'My dog howled; my hens didn't lay; fish jumped out of water; the cows didn't go in the barn'—all right before a quake. You can't draw any hard conclusions from this stuff. But actually I believe many of the stories. They're so universal that it seems at least *some* must be real. A day or two before the small 1977 quake east of Berkeley, some pets acted so strangely that their owners took them to a vet; he said they were very frightened, but he couldn't say why. On the day after the quake, those same animals quieted down and acted normally. Now, this is all a matter of record—but how do you explain it?"

Do animals detect tiny magnetic changes and sounds preceding earthquakes? Probably not. But they may sense the release of many small electric charges that often accompany increased strains within the earth's crust. Whatever the explanation, animal "earthquake behavior" could prove invaluable, for—unlike changes in tilt, compression, and other geophysical measurements that take place weeks, months, or even years before the actual tremor—animals seem to sense quakes within days or even hours. Says Evernden, "They're the only clue we have that short-term earthquake precursors even exist." But which animal stories are true? Seismologists can't predict a quake just because Aunt Minnie's cat refuses to eat. And how many anecdotes do you need to issue a reliable prediction? Are ten enough—or ten thousand?

Ruth Simon, a geophysicist and biologist associated both with the USGS and the Colorado School of Mines, is one of three or four Americans who have tried to answer these questions. She first considered using monkeys for her research, "But they're so complex, it's impossible to tell if they're reacting to something we can't sense, or just hungry."

And so, Simon turned to a much simpler animal, one "less disturbed by external cues"—the cockroach. "I know, I know," she said. "I make jokes about it myself. I *hate* cockroaches—I know what it's like to share an apartment with them. But sometimes even the things we hate have something to offer, and if roaches can help us predict quakes, we ought to take advantage of them."

The cockroach makes an ideal laboratory animal, she explains, because of its elementary nervous system and its unfinicky taste buds (one chunk of dry dog food feeds three roaches for a month). Also, its anatomy and normal behavior patterns are well known. Simon finds it especially intriguing that this scourge of the kitchen has persisted, basically unchanged, for the past 250 million years. "It's ecological success is so great that whatever could kill it—including earthquakes—hasn't."

Her latest experiment was simplicity itself: At two seismic stations on the San Andreas Fault, roaches lived in cages equipped with electronic sensors that responded to the animals' movements. Simon later correlated these records to seismic ones.

"It seems there's increased roach activity up to 36 hours before earthquakes and big storms," she says. "But it's not conclusive yet. And *during* the quake they're very quiet. I've no idea what they respond to, but I know they respond. Perhaps all animals can sense earthquakes to some degree, if we only observe them closely enough."

In the future, Simon hopes to expand her roach network along the San Andreas Fault, and to experiment with snakes and other animals that may react sooner, giving her more lead time for predictions.

It's just possible, she feels, that the world eventually will come to rely on some of its most reviled creatures—cockroaches, snakes, and who knows what—for signals of imminent earthquakes. Fanciful? Perhaps. But then the oft-suggested concept of drifting continents met with ridicule for decades—until only a few years ago, when the plate tectonics theory took hold and totally revised man's view of his earth.

With an "earthquake weathercock," made in A.D. 132, the ancient Chinese detected the tremors that wracked their land. When a quake strikes, a pendulum swings toward the shock's source, jarring a dragon that opens its jaws and drops a ball into a frog's mouth. Throughout history man has searched for ways to understand the awesome forces that shake and rend our earth.

By William R. Gray

Volcanoes: Mountains Born of Fire

"Peelè the Goddess . . . Rolling her anger
Thro' blasted valley and flaring forest
in blood-red cataracts down to the sea!"

—Alfred, Lord Tennyson, *Kapiolani*

WILLIAM R. GRAY, NATIONAL GEOGRAPHIC STAFF

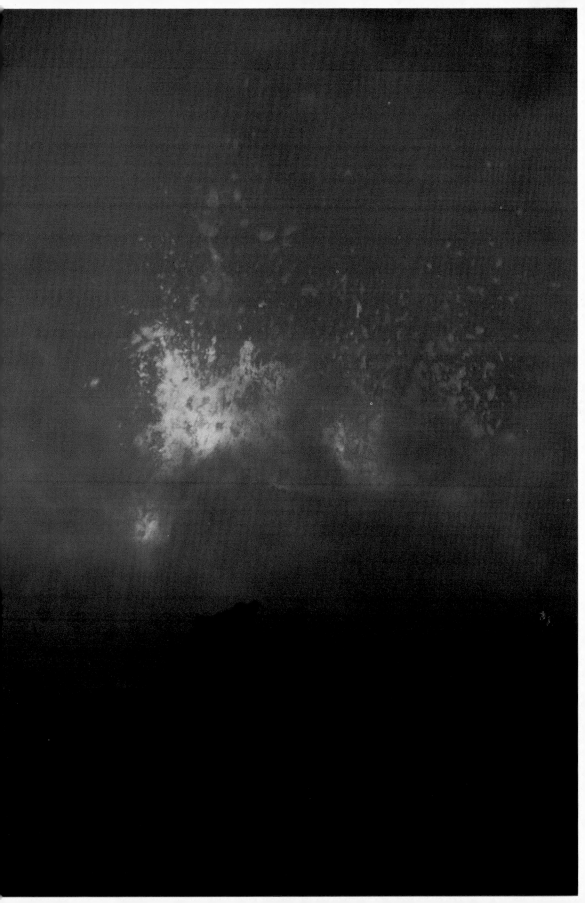

Searing, unrelenting heat blazed from an inferno of molten rock that flooded the night sky with an eerie incandescence. Clouds of vapor—acrid with sulfurous fumes—whirled past in a humid mist. The very earth I stood on groaned and muttered, trembling with tremendous pressures seeking release.

I carefully picked my way through this maelstrom of noise, heat, and light, climbing a live lava flow that was crusted with black cinders. Cracks revealed the orange-red flare of lava—heated to more than 2,000° F.—coursing a few inches beneath my feet. Just 75 yards ahead of me roared the source of this lava—a rent in the earth some 300 feet long that jetted a curtain of molten rock high into the air. The lava rose in pulsating plumes—white-hot at the base, blood-red at the top—that crescendoed in delicate arcs to 400 or more feet before falling, seemingly in slow motion, back to the earth.

Despite the precariousness of my position I stood transfixed, riveted by the majesty and the forbidding power of this phenomenal display of nature's pyrotechnics.

I had flown to the big island of Hawaii in September 1977, summoned by geologists studying the state's volcanoes. An outpouring of lava on the flank of Kilauea volcano promised to build into a spectacular eruption. And build it did. For more than two weeks, lava fountained from deep within the earth, forming meandering flows that burned avenues through a rain forest on their inexorable march toward the coast. One flow crept toward the small village of Kalapana.

Older townspeople there prayed to Pele, fickle goddess of volcanoes; seemingly in answer, the lava stopped at the edge of town. I came to understand the existence of Pele, for I, like the ancient Hawaiians, imagined I could see her face in the clouds roiling above the volcano.

For centuries, Hawaiians found evidence of Pele in every eruption. Small oblong blobs of lava—shaped and cooled by flight through the air—became Pele's tears; others, drawn into fine strands, were Pele's hair. To appease this powerful deity, some Hawaiians built *heiaus*—temples—in her honor at Kilauea, and offered animal sacrifices.

Long ago, an island chieftain named Keoua paid a heavy price for insulting Pele. For weeks Keoua's 400 men had been dutifully making offerings of fish to the goddess, but during a march to meet the army of a rival, they inadvertently offended Pele by rolling stones into her home in the crater. As they crossed the southeast slope of Kilauea, a sudden eruption spread fumes and ash over the area, killing a third of the force.

About two centuries later, while exploring the most recent eruption, I flew over Kilauea by night in a small airplane, circled it by day in a helicopter, hiked around its perimeter, slept next to its warmth, dried rain-soaked clothes by its radiant heat, and walked on its lightly crusted rivers of fire.

On the flank of this volcanic giant, I felt the troubled pulse of a restless earth.

"To me, a volcano in eruption is a mountain gone mad," Dr. Robert W. Decker told me. "They're wild, violent, and largely unpredictable." A robust man with dark curly hair and a broad smile, Bob Decker is a highly respected earth scientist. I visited him on a wintry day in his office at Dartmouth College in New Hampshire to learn more about the nature and origins of volcanoes.

"Like most other fields of geology, volcanology was revolutionized with the advent of the theory of plate tectonics," he began. "We now see volcanoes as occurring in three general areas, each related to the plates."

In the first situation, Bob explained, volcanoes appear where two plates are pulling apart. Magma, or molten rock, wells up from far below the earth's surface and spills out, usually on the ocean floor, as lava. Mountains are built, and some may even break the surface. The Mid-Atlantic Ridge, for instance, exists along the seam where the Eurasian and American plates begin to move in opposite directions. It is a chain of mountains that reaches above sea level in such places as Iceland, the Azores, and Tristan da Cunha.

Volcanoes are also born where plates collide. As one plate slides under another, it melts and churns as it descends. Molten material rises to the surface and erupts there, forming new land. Through such collisions, the Caribbean Islands were formed, the volcanic peaks of Central America grew, and the lofty, glacier-clad mountains of the Cascade Range in the Pacific Northwest were built up.

"Both of these kinds of volcanoes occur along plate margins — the boundaries between tectonic plates," Bob continued.

"The final type of volcano appears in the middle of a plate. Somehow, a hot spot in the earth's mantle melts a hole through the middle of a plate, allowing molten material to spill out on the surface. The Hawaiian Islands are a perfect example of this process — they are right in the middle of the huge Pacific plate."

Bob settled back in his chair and glanced out the window. Snow clung to the branches of bare trees, sharply etched against a dull gray sky. "You'll learn to love the beauty of volcanoes," he told me. "Despite their violence, they have a grace and nobility unmatched by other mountains. And each one seems to have a personality of its own."

During the next few months, as I traveled widely to many volcanic areas around the world, Bob's words returned to me often, and I began to develop definite sensitivities to certain mountains. Some seemed haughty, others enticing; some harsh, others gentle. But each had a certain beauty that stemmed from (Continued on page 56)

Struggling to save his home, an Icelander shovels ash from the roof to keep it from collapsing. In 1973 a volcanic eruption threatened to bury his town, Vestmannaeyjar.

Preceding pages: Plumes of lava jet from a rift on the flank of Kilauea, a volcano on the island of Hawaii. Geologist Jack Lockwood dashes over the hot crust to take samples of molten rock for clues to the geologic history of the volcano.

Preceding pages: Rain of lava bombs and ash build Eldfell— "Fire Mountain"—where only a plain existed before. Snow-frosted ash blankets Vestmannaeyjar on the island of Heimaey. The fireworks began on January 23, 1973; within five months Eldfell had grown to 705 feet.

Searing wall of lava bulldozes a house in Vestmannaeyjar and crumples corrugated sheetmetal. Townspeople had used the metal in a futile effort to protect their windows. Though lava, ash, and fire destroyed a third of the town's 1,200 homes, all 5,000 inhabitants escaped injury, and most sought refuge on the

PRECEDING PAGES: NATIONAL GEOGRAPHIC PHOTOGRAPHER EMORY KRISTOF. SIRGURGER JONASSON

mainland. Weeks later, a resident entered a cellar and died, overcome by toxic gases trapped there. When the eruption ended, ash lay five feet deep, burying the base of a gravesite Madonna (right). Four years later, after a clean-up campaign by the resolute islanders, the statue gleams—and the town stands rebuilt.

ROBERT S. PATTON, NATIONAL GEOGRAPHIC STAFF

the dramatic fire-power of its origins.

Iceland, an island-country half the size of Kansas, is one of the world's most volcanically active places. It straddles the Mid-Atlantic Ridge, and during its eleven hundred years of recorded history it has been hit by an eruption on the average of once every five years.

"Basically, Iceland is part of the ocean floor," said Dr. Sigurdur Thórarinsson, the dean of Icelandic geology. "We simply happen to be sitting over a plume where more magma has been released and therefore more terrain built." I met Dr. Thórarinsson, a slender man with graying hair, in his office at the Science Institute in Reykjavík, Iceland's capital. "Because we are on the rift between two plates, Iceland is actually widening—the eastern part of the country is moving toward Europe and the western part toward North America. The rate of widening is small—just half an inch or so a year on the average—and intermittent."

The next morning, following Dr. Thórarinsson's advice, I drove to Thingvellir—a ragged slash in the earth ten miles long. Although it was well past ten o'clock, the sun was just rising—a February dawn comes late to Iceland, whose northern coast grazes the Arctic Circle.

Parallel walls of black basalt dropped steeply to an uneven floor. I scrambled down one and immediately felt the numbing chill of an Icelandic winter; there was no snow, but a hard frost gripped the land.

From the bottom of this 70-foot-deep chasm, I reflected that I was standing at the basement of the earth's surface, the seam in the foundation below which existed the churning mass of the mantle.

The next day, from an airplane, I could see the relationship between the rift zone and its associated volcanoes. Several large volcanoes rose within a few miles of Thingvellir, and long flows of lava—virtually untouched by vegetation in this harsh climate—carpeted the ground.

A 15-mile-long fissure pocked with craters and vents marked the site of Laki, where—in terms of lava

produced—the largest eruption in recorded history occurred. In June 1783, floods of basalt poured from Laki and flowed more than 35 miles.

The actual eruption caused little damage, since Iceland was sparsely populated. Its long-term effects were disastrous, however. During the summer a light bluish haze, caused by the volcanic gases, drifted over not only Iceland but also Europe

and adjacent parts of Asia. In Iceland, the gases stunted the grass crop and destroyed grazing areas; about 70 percent of Iceland's livestock died during the next two years. Wholly dependent on raising sheep and cattle, Iceland's people soon began to suffer. In all, some 10,000 people — one-fifth of the population — died of starvation. Years passed before the country recovered. *(Continued on page 62)*

Battling advancing lava that threatens to close their vital harbor, Vestmannaeyjar workers cool and harden the leading edge of the advancing flow by spraying it with seawater. The lava dam diverted the molten rock behind it, creating a breakwater that actually improved the harbor, giving it greater protection from the sea.

Fiery Messengers From Earth's Churning Mantle

Born of the restless shifting of earth's tectonic plates, volcanoes erupt when magma from the interior reaches the surface. Generally, three basic types of volcanoes occur. Draining the ocean beds of water reveals the titanic forces at work.

Mid-plate volcanoes, at left, such as those of Hawaii, lie over hot spots in the mantle; plumes of magma melt openings in the crust, and volcanoes erupt. The plate drifts

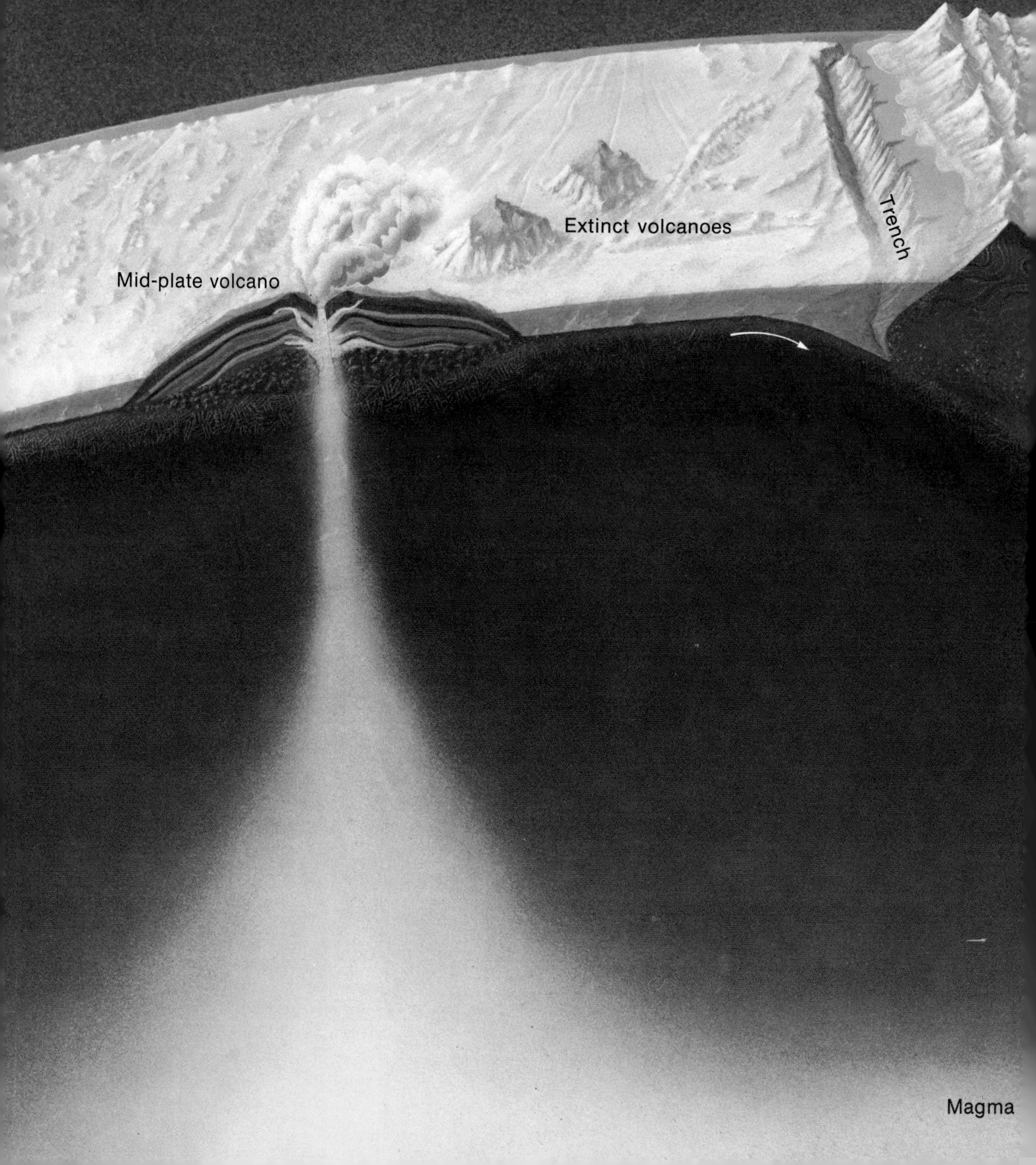

Extinct volcanoes

Trench

Mid-plate volcano

Magma

over the hot spot, creating a row of volcanoes; only the most recent ones remain active. Mid-plate volcanoes also puncture continents, in places such as northern Africa.

In the center, a plate underlying an ocean dips beneath one carrying a continent, creating a deep trench such as that along the Pacific Coast of South America. The plates grind across each other, building tremendous heat and pressure and melting rock to form magma that erupts in a volcano. Some 80 percent of active volcanoes occur along such plate margins.

At right, a volcano erupts where two oceanic plates separate. Lava breaks through the line of weakness, or rift, gradually building mountains. Sometimes the volcanoes reach the surface of the sea and become islands—such as Iceland on the Mid-Atlantic Ridge.

PAINTING BY JAIME QUINTERO

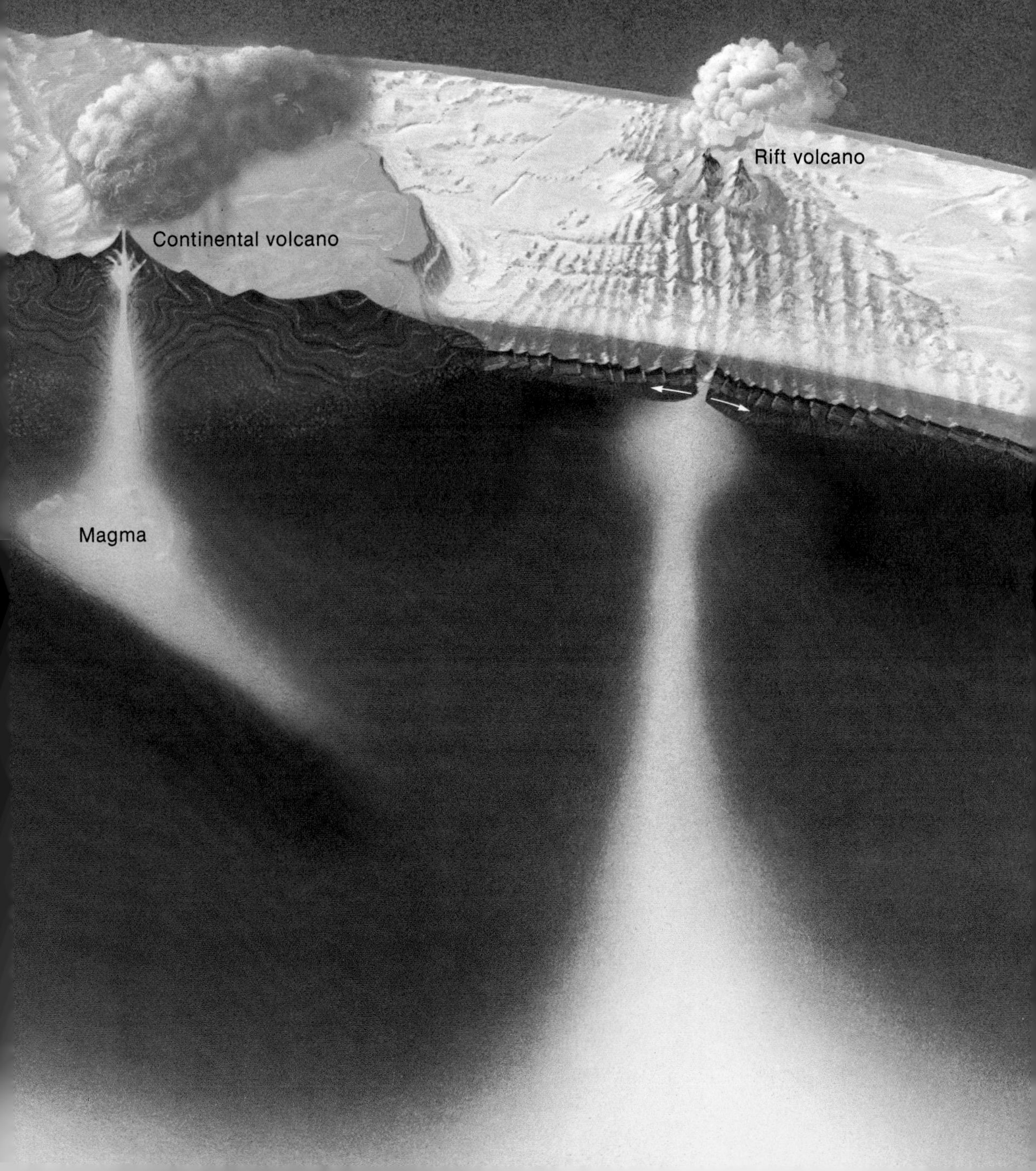

Rift volcano

Continental volcano

Magma

Twin summit craters crown Europe's highest volcano—Mount Etna in Sicily. A still-active volcano that rises 10,902 feet, Etna has had recorded eruptions since 500 B.C. Continuous rumbling and smoking in ancient times fostered myths of Vulcan, the Roman fire god, working his forge deep within the mountain, and of the one-eyed giant Cyclops raving there.

Real-estate developments now
cover lava flows barely 70 years
old. Villagers (above, left)
view lava streaming down Etna in
the late 19th century. During
eruptions in 1971 and 1974, police
had to chase away sightseers who
pressed too close for safety.
Sprawled plaster figures at
Pompeii, in Italy, tell of sudden
and terrifying death (left). In A.D.

79 Mount Vesuvius erupted,
covering the city with a fiery cloud
that suffocated thousands of people;
volcanic ash buried many of them
20 feet deep. Archeologists digging
into the ruins of Pompeii found the
holes left by decayed bodies; in a
technique used since 1860, they
painstakingly pour plaster into the
molds, creating eloquent casts of
the volcano's victims.

*Reduced to rubble, St. Pierre on the
Caribbean island of Martinique
reveals a volcano's enormous power.
On May 8, 1902, Mont Pelée
sent a superheated cloud of gases
rolling down its flank. In seconds, the cloud reached St. Pierre,
killing all but one of its 30,000
inhabitants. Days later, a visitor
(lower) views the devastated city.*

Scattered bits of rock stringing southward from Iceland form the Westmann Islands. Surtsey, one of these, was created in 1963 when a submarine volcano reached sea level and built an island 570 feet high. Just north of Surtsey, on the island of Heimaey, a new volcano now called Eldfell began erupting in 1973. It threatened the town of Vestmannaeyjar, whose residents fought the encroaching lava with water pumped from the Atlantic. This water cooled and hardened the leading edge of the lava, diverting the flow. Although one-third of the town was eventually swallowed up, the efforts of the people of Vestmannaeyjar saved most of their homes and heritage.

Other people have not fared so well. The city of St. Pierre on the Caribbean island of Martinique was incinerated almost instantly on May 8, 1902. Its inhabitants—more than 30,000 people—were killed. Mont Pelée, today a green-shouldered mountain draped with mists, erupted with an incandescent mass of gases and ash called a *nuée ardente*, or glowing cloud. This billowing, super-heated mass rushed down the slopes at a hundred miles an hour. It struck the town and left nothing un-damaged. "There was no warning," wrote an officer aboard the *Roraima*, a Canadian ship anchored in the harbor. "The side of the volcano was ripped out, and there hurled straight toward us a solid wall of flame. It sounded like thousands of cannon. The wave of fire was on us and over us like a lightning flash. It was like a hurricane of fire.... The town vanished before our eyes and then the air grew stifling hot and we were in the thick of it. Wherever the mass of fire struck the sea the water boiled and sent up vast clouds of steam.... I saved my life by running to my state-room and burying myself in the bed-ding. The blast of fire...shriveled and set fire to everything it touched. Burning rum ran in streams down every street and out onto the sea. This blazing rum set fire to the *Roraima* several times. Before the volcano burst the landings of St. Pierre were crowded with people. After the explosion not one living being was seen on land."

In fact, only one person in town survived—a prisoner locked in a thick-walled cell. Even today, St. Pierre bears witness to the ravages of the nuée ardente—bare building foundations, piles of rubble where houses once stood, tombstones knocked flat in a graveyard.

"The disastrous eruption of Mont Pelée is typical of volcanoes in the Caribbean—and indeed typical of all areas where two tectonic plates collide," said John Tomblin, British-born geologist who heads the Seismic Research Unit at the University of the West Indies in Trinidad. "The Caribbean plate is like a long tongue overriding part of the American plate, which originated in the mid-Atlantic. The Caribbean's movement is eastward and upward relative to the American, and the island arc of volcanoes that forms the Lesser Antilles rises just west of the juncture. From Grenada on the south to Saba on the north, these islands have shown volcanic activity within recent geologic time—the last ten thousand years. Eruptions add a little to the size of the islands."

I flew north to the tiny island of St. Vincent to see some of the most recently formed Caribbean land. On a sultry, windless morning, I climbed a volcano named La Soufrière—the French word soufre, or brimstone, describes many active areas in this region—on a trail that aimed straight to the top, some 4,048 feet above the shoreline. I was accompanied by Edgar Williams, a sturdy, compact native of St. Vincent who scales the volcano each week to make scientific measurements. He set a pace which, in that heat and humidity, I found difficult to match—but when I reached the summit I was rewarded.

I stood at the rim of a large crater whose interior walls—decked with light-green moss—plummeted to a lake gently rippled by a soft breeze.

Contrasting markedly with the serenity and soft beauty of the lake was an island of naked, steaming lava. In 1971 the temperature of the lake began to rise, and soon a jumbled mass of fuming rock broke the surface. It gradually built an island 216 feet high that filled one-third of the lake. This lava dome still emanates sulfurous fumes, and hot spots discolor the water around the island. Edgar measured the temperature of the water at 78° F.—fully one hundred degrees lower than the 1971 high that cooked every fish in the lake.

I sat at the water's edge chatting with Edgar and chewing fresh sweet sugarcane that he had cut on the way up. I recalled a lunch I'd had at the edge of another lake within a volcano. Crater Lake in Oregon spreads midnight blue water within the remnants of a once-towering volcano named Mount Mazama. Until 6,600 years ago, this volcano stood at 12,000 feet. A series of enormous eruptions emptied its magma chamber—the subsurface reservoir of gas and molten rock that feeds the volcano—and the mountain collapsed in on itself to half its height, forming a caldera. Gradually, rain and snowmelt filled the depression to form Crater Lake, which measures six miles by four and plunges to 1,932 feet—the deepest lake in the U. S.

Mount Mazama is but one in the chain of elegant, skyscraping volcanoes that crown the Cascade Range, running from Lassen Peak in northern California to Mount Baker in Washington. Created by the melting of the Pacific plate as it slowly descends beneath the North American plate, the Cascades form part of the "ring of fire"—that girdle of tremendously active volcanoes circling the Pacific Ocean.

To me, the undisputed queen of the Cascades is Mount Rainier—a glacier-draped giant that rises to 14,410 feet, nearly two miles higher than the surrounding land. Naturalist John Muir considered Rainier "the most majestic solitary mountain I had ever beheld."

A long, steep ascent up rumpled glaciers, across plunging crevasses, and past jumbled icefalls brought me to Columbia Crest—the high point on the crater rim. Below me, an almost perfectly symmetrical crater half a mile across was filled with ice and snow. Despite the thin air, the cold, and the shroud of glacial ice, Mount Rainier is still a hot mountain. Jets of steam hissed from the rocks throughout the crater. This heat has carved labyrinthine passages of steamy, chill blackness under the ice.

Despite Rainier's sublime countenance—Indians called it Tahoma, the mountain that was God—its volcanic activity has been marked by violence. And that violence almost certainly will continue. The greatest danger, apart from a climax eruption like Mount Mazama's, is from mudflows. A thick slurry of earth, rock, and other material, a mudflow is triggered by the slipping and sliding of loose rock that has been saturated with water, or by eruptions that melt large masses of ice and snow. These flows, often dozens of feet thick, thunder down river valleys at more than a mile a minute. Several towns north and west of Rainier are built on ancient mudflows and could be engulfed by future ones. Also, towns to the east of the mountain—the direction of the prevailing winds—could be blanketed by ashfall.

While towns in the Pacific Northwest exist under the potential threat of the Cascades volcanoes, communities on the big island of Hawaii actually lie in the shadow of two of the most active volcanoes in the world—Mauna Loa and Kilauea. At least one settlement has been destroyed by lava in recent decades, and Hilo, the island's largest town, lies on the path taken by some of the longest lava flows on Hawaii.

The two active peaks are simply the current manifestations of a long chain of volcanoes that stretches across the Pacific Ocean for 3,300 miles. "The existence of Hawaii," explained Dr. Gordon Eaton, the scientist in charge of the U. S. Geological Survey's Hawaiian Volcano Observatory, "is due to a hot spot or volcanic plume that has cut through the Pacific plate like a welder's torch through metal. This plume has kept a relatively constant position during the last tens of millions of years while the Pacific plate has inched northwestward over the top of it.

"As the plate marched over the plume, a series of volcanoes was built beginning near the west end of the Aleutian Islands and ending, at least for now, here on Hawaii. Some of these volcanoes never broke surface; others did and were subsequently eroded away. Others form the base for coral atolls, while the younger ones—the main islands of the Hawaiian chain—remain as true volcanic islands. We've found evidence that may indicate the plume is building new volcanoes southeast of Hawaii. Perhaps in the future there will be another Hawaiian island."

I investigated the volcanoes on four of the major islands and found a diversity of landforms. But only on the island of Hawaii do the volcanoes of the 50th State reach truly sublime majesty. Mauna Loa gracefully lifts its head 13,677 feet above the deep-blue waters of the Pacific. It rises gently, deceptively, looking more like a smooth-sided hill than one of the grand mountains of the world. But indeed it is. From its base on the seafloor to its lava-draped summit, Mauna Loa measures more than 31,000 feet, surmounting even Mount Everest, at 29,028 feet the tallest mountain on land.

Staggering, too, is the manner in which it gained its height. About a million years ago, a crack opened at the bottom of the sea, and a puddle of lava a few feet thick spilled out,

(Continued on page 70)

Volcanologist Robin Holcomb flinches from the heat of an incandescent lava fountain on Mauna Ulu. The vent, on the flank of Kilauea, erupted periodically from 1969 to 1974. The pressure of escaping gases powers the pulsing fountain. Holcomb—of the Hawaiian Volcano Observatory—collects fresh spatter to study the mineralogical and chemical composition of the lava.

ROBIN T. HOLCOMB, HAWAIIAN VOLCANO OBSERVATORY (ABOVE AND BELOW)

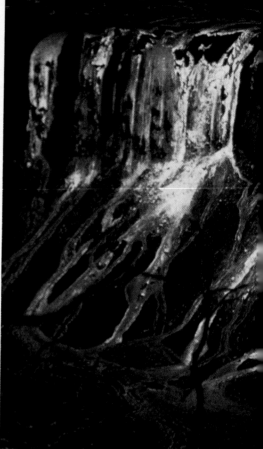

Curtain of fire (below) jets from a fissure on Mauna Loa on the island of Hawaii. The world's largest mountain in volume—some 10,000 cubic miles—Mauna Loa erupted in 1975 after remaining dormant for 25 years. At right, lava cascades over the rim of a crater near Mauna Ulu. The fluid lava of Hawaiian volcanoes readily releases gases. This creates spectacular fireworks, and pressure within the volcanoes seldom builds to explosive force. Above, lava bursts from the top of Mauna Ulu in 1974.

DONALD A. SWANSON, HAWAIIAN VOLCANO OBSERVATORY

Beside a fuming fissure in the caldera of Kilauea, chemist Bruce Finlayson takes gas samples after an eruption in 1971. Having monitored gases, earthquakes, and other cycles of activity on Kilauea for 65 years, scientists often can anticipate its major eruptions.

Scientists use special instruments to investigate the inner workings of volcanoes. At left, a geologist surveys a volcano on Hawaii. A Geodimeter in the distance emits a laser beam that strikes the reflector on the tripod and returns. The time lag yields a measurement of the distance. Repeated readings detect internal swelling in the volcano, thus hinting at an eruption. Geophysicist Len Anderson studies changes in a volcano's electromagnetic field. Shifts in intensity and direction of the field sometimes indicate movement of magma at shallow depths.

SAM ABELL

Meandering seaward, a fiery river burns through a rain forest on Hawaii in 1974. At the hotter center of the river, the lava moves several feet a second. Hawaiian volcanoes sometimes produce a thin, fluid lava that may flow as far as 35 miles.

Smooth skin of pahoehoe lava piles up in ropy, yard-wide pleats as it oozes from a vent on Mauna Ulu in 1973. Pahoehoe flows faster than rough, blocky aa lava. Pahoehoe and aa may have the same chemical makeup, but differ in their physical properties when molten. A moving pahoehoe flow may change into aa.

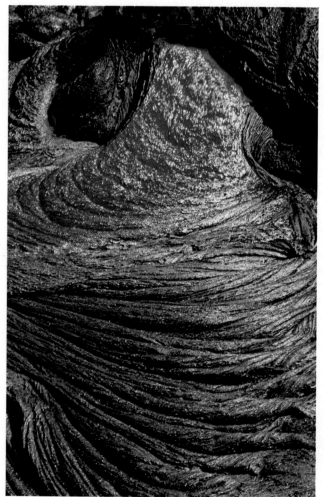

cooling quickly in the frigid depths. Slowly, lava flow piled on lava flow until the edifice of today was built.

But Mauna Loa's fires are far from banked, and the mountain is still growing. A 1975 eruption added new lava, and more eruptions are expected in the future.

Just past dawn on a crisp morning, I awoke high on the northeast flank of Mauna Loa. The afternoon before, I had driven with Keith Hoofnagle, naturalist at the Hawaii Volcanoes National Park, to a small

weather observatory 11,000 feet up. We slept there, allowing our bodies to grow accustomed to the thin air before we climbed to the summit.

Shouldering our packs, we started out with a sun that was already baking the black lava we walked on. Keith, a tall lanky man with a gentle temperament, led the way. No true trail existed, but a series of yellow marks painted on the *pahoehoe* (pa-HOY-hoy) lava indicated a general course uphill. Pahoehoe is blackish-gray, solid, fairly smooth, and easy to walk on. When fresh and flowing, it moves with little difficulty over the land, following natural contours. Within the first mile we crossed a flow of *aa* (ah-ah) lava, and I quickly learned the difference between the two. Aa is sharp-edged, chunky, and difficult to walk on. It shifts, crumbles, and quickly chews up hiking boots. "Pahoehoe has a higher gas content," Keith said. "It thus flows much more rapidly and cools into smoother rock."

The wind, gentle all morning,

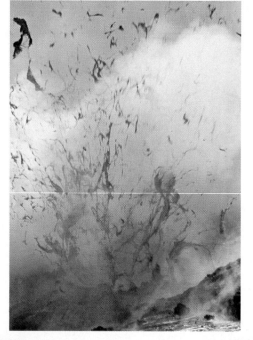

Snake of fire from Mauna Ulu slithers across smoldering lava that has covered a highway on Hawaii. Eruptions from 1969 to 1974 buried 12 miles of road up to 300 feet deep. The youngest mountain in the United States, Mauna Ulu—"the growing mountain"—poured enough lava into the ocean to add some 200 acres to the island of Hawaii. Steam and lava explode (left) as an ocean wave washes over a flow seven miles from the vent. Despite frequent eruptions, Hawaiian volcanoes take few lives. The last fatality occurred in 1924.

Lava from Mauna Ulu dogs Robin Holcomb as it advances through a forest. Plants decompose beneath lava flows, producing gases which seep through the porous ground; when they surface ahead of a flow, heat from the lava ignites them.

72

began to freshen, and soon we were enveloped in a thick mist that cut visibility to a few hundred feet. We worked our way through the mist until, after four hours of climbing, we neared Mokuaweoweo, the huge caldera of Mauna Loa.

We skirted north of the caldera, crossing alternate flows of pahoehoe and aa, until we abruptly descended to the floor of the North Pit, a small crater adjoining the larger one. In the mist, we followed rock cairns that marked the trail across this half-mile-wide crater that was covered with crinkly black lava from the eruption of 1975.

The last 500 feet of elevation to our destination—a cabin on the caldera rim—covered a mile and a half of aa lava. As I trudged uphill, panting now in the rarefied air, I felt the temperature plummeting. Almost immediately a porous hail began to fall that soon turned to a heavy snow.

We reached the cabin within a few minutes and escaped the silent fury of the snowstorm—but not the cold. Inside the cabin—a bare two-room structure with a few bunks—I felt even colder, and my breath formed dense clouds. Pulling on a wool shirt, down parka, hat, and mittens, I began heating water for tea.

Half an hour before sunset, the snow finally stopped. The clouds began to break apart, and shafts of sunlight broke through, illuminating Mokuaweoweo for the first time. I walked the ten yards from the cabin to the caldera rim and was awestruck. A huge gash in the earth three miles by one and a half opened before me. Its floor, 450 feet below, was rumpled with small cones and swept with lava flows. A bluish smoke wafted from a long crack, marking the fissure of the 1975 eruption. "There's still some gas and enough heat to turn water to steam," Keith remarked.

That night, the snow returned, and the ground was thickly blanketed by dawn. We hurriedly packed and found our way down the mountain. The next day, after we had returned to the national park headquarters on Kilauea, I glanced up and saw a crown of white on Mauna Loa.

The snowstorm had departed, but had left its pristine legacy.

From a distance, Kilauea seems like Mauna Loa's little sister; it is a full 9,500 feet lower in elevation. But, if anything, it is more active than the larger volcano, having erupted 29 times in the last 25 years. Its main focus of activity is a large caldera with a fire pit called Halemaumau sunk into its floor.

American missionary William Ellis was the first outsider to see this pit. He recorded his impressions in 1823: "Astonishment and awe for some moments rendered us mute, and, like statues, we stood fixed to the spot, with our eyes riveted to the abyss below. The bottom was covered with lava, and the southwest and northern part of it were one vast flood of burning matter, in a state of terrific ebullition, rolling to and fro its 'fiery surge' and flaming billows."

Even when it's not in eruption, Kilauea fumes and sputters, exuding gas and steam with a distinct aroma. "The smell of sulfur is strong," observed Mark Twain, "but not unpleasant to a sinner."

Kilauea's eruptions often give birth to new land, as lava pushes into the sea and creates a maelstrom of steam. As the lava cools, the sea begins its relentless erosion, carving cliffs and sea arches. Hawaii's famous black beaches were formed when seawater quenched hot lava as it poured into the ocean, and pulverized it in explosions of steam. At Kaimu, on Hawaii's south shore, I found such a beach sheltered by palms. The black seemed out of place—a startling contrast to the foaming white surf and the vivid green of the palms.

I explored another spot of exquisite beauty in the realm of Hawaii's volcanoes named Kipuka Ki. A *kipuka* is a chunk of land that has been surrounded by lava flows—an island of established vegetation in the midst of desolation. Often these isolated bits of land have become distinct ecosystems, harboring plants, insects, and animals unique to that kipuka. In Kipuka Ki, an enclave of just a few dozen acres, contorted koa trees form a green canopy above a garden of luxuriant ferns. Birds fluttered among the branches, making the only sound in this gentle wilderness nurtured by volcanoes.

Absolute blackness confronted me in another volcanic wilderness. Sitting on a smooth hump of lava, I raised my hand until it touched my nose — and saw not so much as a flicker of motion. I was exploring Arco Tunnel in southern Idaho, and had lagged behind my companions to experience the darkness. I watched their lights bob out of sight and then let my vision adjust. The blackness lay like a shroud.

Arco Tunnel is one of dozens of lava tubes in Craters of the Moon National Monument, an 83-square-mile belt of cinder cones and lava flows that ceased erupting just two thousand years ago. I climbed into the tunnel's mouth with Ken Harrison, park naturalist, and Bill Sidle, a geology student. "Tubes like Arco Tunnel are created during an eruption that produces lava streams that flow for a long time," Bill explained. "The air cools the top layer of molten material, forming a skin of hardening rock. The hot interior of the flow con-

---

"Nature understands no jesting. She is always right, and the errors and faults are always those of man."

— GOETHE

---

tinues to move, however. Eventually, when the eruption ceases, the lava drains out, and a tube remains."

A markedly imperfect tube, I discovered. At one point we had to slither through an opening barely a foot high. In another place, the tube opened into a spacious room with a ceiling 20 feet in height. Generally, though, the tunnel was about six feet high — I could tell because I kept bumping my hard hat on small stalactites. These stalactites were formed when drips of hot lava cooled and became rock as the lava tube emptied of molten material. They

appeared at times in thick clusters. In one place, droplets of water decorated the tips, and in the beam of my light gleamed like diadems.

In Craters of the Moon, I encountered other phenomena typical of volcanic areas. On the steep flanks of Big Cinder Butte — the highest of the two dozen cinder cones in the monument — I found a crescent-shaped lava bomb some two feet long. Blown from the crater as a fiery-hot chunk of lava, it had twisted and cooled during its flight. The resulting rock looked like saltwater taffy that had been pulled on a machine. In another part of the monument, I found the remains of a forest that was flooded by a pahoehoe flow. The pasty lava had wrapped itself around the trees and formed casts as it cooled. Some of the casts retained the shapes of the standing trees; smaller ones showed impressions of charred wood.

Just a hundred miles east of Craters of the Moon sprawls one of the world's most unique, most captivating, and most visited regions — and yet its geologic history has been deciphered only during the last few years. Yellowstone National Park encompasses a wonderland of forests, mountains, rivers, waterfalls, and spectacular geysers and hot springs. But its essential character is volcanic. To discover the dynamic processes behind the development of this magnificent national park, I arranged to spend several days with Rich Reynolds, a 30-year-old geologist with the U. S. Geological Survey. Rich had worked five summers with Robert Christiansen, the geologist who developed and proved the theories behind Yellowstone's history.

"Volcanism has been a part of Yellowstone's geology for millions of years," Rich began. "The Absaroka Range over there" — he pointed to craggy gray mountains in the east — "consists largely of volcanoes probably 50 million years old. But the recent major activity, and the main designer of Yellowstone's landforms, was a series of ash-flow eruptions during the last two million years. We'll concentrate on one that occurred 600,000 years ago."

Molten rock glows through the collapsed crust of a lava tube on Mauna Ulu. Geologist Donald Swanson, treading softly, carries a surveyor's rod; he studies volcanoes before and after eruptions to record their changing shapes.

Ill-tempered goddess of volcanoes in Hawaiian legend, Pele haunts the fiery pit of Kilauea. According to author Will Gray, "It's easy to imagine many fantastic images — including Pele's face — in the roiling clouds of a volcanic eruption." Volcanoes sometimes spray tear-shaped drops of lava; ancient Hawaiians named them "Pele's tears." Long, thin strands they called "Pele's hair."

During the next few days, we explored the geology of Yellowstone: We climbed sagebrush-strewn ridges to see the contact between two ash flows; we explored conifer-covered lava flows; we rested in meadows fired by scarlet Indian paintbrush and lavender lupine; and we hiked into a wilderness geyser basin.

During this time, I learned of the cataclysmic ash-flow eruption. "There was probably a gradual swelling as the hot material slowly rose to the surface. Then a series of fantastic eruptions emptied the magma chamber. Huge ash flows gushed from the vents and roared down river valleys at incredible speeds. Windblown particles have been traced clear to western Iowa and to California. These were rapid and destructive eruptions; during our time on earth, we have seen nothing like them.

"After the ash flows, the ground above the magma chamber subsided, forming a caldera that measures 45 miles by 75 miles. It covers most of the south-central part of the park and includes Yellowstone Lake. This massive caldera has been partly filled by lava that flowed from cracks and openings in its floor. The lava was thick and viscous and formed mounds and plateaus rather than long tongues as in Hawaii. Most of the ridges you see around the park are forest-covered flows, and the most recent is only about 70,000 years old."

Puzzled, I asked Rich why a volcanic center existed here at all, hundreds of miles from any plate margin.

"We think, but we're far from sure, that the collision of the North American and Pacific plates produces stresses in the North American plate that tend to pull apart large regions of the western U. S. An ancient deep-seated fracture extends along the East Snake River Plain at least to Yellowstone. Melting at the base of the plate may be augmented by the stress along the fracture. Magma thus has an avenue through which to rise to the surface."

Another theory is that Yellowstone is a hot spot, like Hawaii, and that the eastern part of the Snake River Plain is a volcanic track left behind as the American plate moved west over the plume.

The geysers, hot springs, and fumaroles of Yellowstone are surface indicators of the existence of magma. To explore a geothermal area, Rich and I hiked through the lush meadows of DeLacy Creek, along the shore of sparkling Shoshone Lake, to the puffing, gurgling, hissing realm of Shoshone Geyser Basin. Here, dozens of geysers and springs issue steam and hot water. Water trickles from a small crack in some; others eject steam in a pulsing cycle high into the air. Gentle Shoshone Creek meanders through this raw wilderness, collecting the flow from the springs. In places, the cold mountain water of Shoshone Creek mingles with the boiling offering of the geyser basin to form pools of pleasant warmth. As Rich and I basked in one of these, I learned something of the mechanics of a geyser basin.

"Groundwater percolates far into the earth," Rich said, "and becomes heated. The water is soon well over the boiling point and circulates upward to find release. As it nears the surface it may turn to steam or remain liquid. Depending on the pressure, it may shoot skyward as a geyser, or simply flow out as hot water."

In Yellowstone I saw the results of a rapid and catastrophic eruption that formed a caldera. In northern New Mexico, on land held sacred by Navajo Indians, I found a volcano that suffered a completely different fate. Erosion by wind and water over

Steam billows from hot springs in the El Tatio geyser field high in the Andes of Chile. Dormant volcanoes rim the area, where hot rock deep underground warms water, forcing it to the surface. Chile hopes to build a geothermal power station at El Tatio, and has constructed a prototype plant that produces fresh water from the geyser steam.

NATIONAL GEOGRAPHIC PHOTOGRAPHER GEORGE F. MOBLEY

Holiday crowd of Icelanders soaks in a thermal pool on a chill summer day in the capital city of Reykjavík. Situated near a volcanic area, Reykjavík taps groundwater for heat. A concrete conduit (above) carries hot water from wells to homes and businesses. Iceland pioneered the use of natural hot water for heating in the 1930's. This clean and inexpensive resource now heats most of the homes on the island at a cost about 75 percent lower than oil. Today the high cost of importing fuel and food spurs even greater use. A gardener (right) gathers tomatoes in a greenhouse warmed by pipes carrying natural hot water. Geothermally heated greenhouses supply fruit, vegetables, and flowers to Reykjavík markets the year round.

WILLIAM R. GRAY, NATIONAL GEOGRAPHIC STAFF (ABOVE). ED COOPER

millions of years has worn a large mountain down to a nub—today, only its innermost core of volcanic rock stands exposed. Ship Rock, its prow rearing 1,600 feet above the desert, is part of the volcanic conduit that fed molten rock from the magma cham- ber to the surface before becoming plugged. Dikes—lava that once filled cracks in the rock and was later ex- posed by erosion—radiate from the

monolith like spokes from the hub of a wagon wheel.

Early one morning, I watched the ever-shifting light of a desert dawn create patterns and colors on Ship Rock that constantly trans- formed its moods. I understood why the Navajo nation once considered it hallowed, and gave it the ethereal name *tse bit'a'i*—"rock with wings."

Volcanoes have always drawn

the reverence of people who live around them. Some cultures considered them actual deities; others thought them to be the dwelling place of the gods. The Incas of South America scaled 20,000-foot ice-clad volcanoes to offer human sacrifices.

Besides evoking the spiritual, however, volcanoes provide many practical benefits to mankind. Ash and lava from their eruptions renew the soil with needed minerals—particularly potassium—fundamental to plant growth. The rich coffee that makes Guatemala famous, for instance, grows best on the slopes of its many beautiful, cone-shaped volcanoes. And the rice crops that sustain multitudes of people in Japan, the Philippines, and Indonesia flourish in the volcanic soils of those islands.

Even the rocks themselves are

useful. Some pumice-rich deposits are extensively quarried for use in soaps, concrete, insulation, acoustic tile, plaster, and metal and furniture polishes. Obsidian, a shiny black glass found in some lavas, was traded by Indians throughout the Americas for use as spear points and arrowheads. Bits of obsidian from Yellowstone have been found as far away as Florida.

But the most treasured of volcanic by-products are the precious metals and gems. The gold and silver rushes of the American West were staged in ancient volcanic areas where hot solutions deposited veins of metal that were later exposed by erosion. Copper, sulfur, and iron ore are also found in volcanic deposits, and diamonds are created in fiery-hot caldrons of magmatic material deep in the mantle of the earth.

Today, however, one of the great potential contributions of volcanoes is geothermal energy. With the worldwide depletion of fossil fuels, geothermal energy — natural heat, steam, and hot water — seems an enticing alternative. But the problem, as with most forms of energy, is how to harness and use it.

A week-long eruption of Kilauea produces enough energy to supply 40 percent of the power demands of the United States for those seven days. But finding a way to tap a volcano in eruption is almost beyond the realm of imagination. Using the nearly continuous source of energy formed by a subterranean body of magma is not, however. Many countries, among them Italy, New Zealand, the United States, and particularly Iceland, have had some success in developing geothermal power, and the technology is rapidly expanding.

For decades, tiny Iceland has led the world in the research and development of geothermal energy. Isolated in the storm-tossed North Atlantic, Iceland long depended on costly shipments of oil for its heat. Today, according to Dr. Gudmundur Pálmason, head of the Geothermal Department of the National Energy Authority, 65 to 70 percent of all homes in Iceland — and all of Reykjavík's buildings — are heated by hot water pumped from the ground. "Soon," he added, "we hope to increase that figure to 80 percent. It is our official government policy to eliminate reliance on oil imports for house heating."

I talked to Dr. Pálmason in his office in downtown Reykjavík — the only major city in the world that is heated by geothermal methods. "The main extension of the Reykjavík system was in the early 1940's," he said. "Before then, only scattered homes were heated with natural hot water. Today, our main source is about ten miles east of the city. The hot water is pumped from the ground and channeled through pipes to the city's distribution system. It averages about 165-175° F. when it reaches the homes. The same hot water that provides the heat is also used for washing, bathing, and even for making coffee — there is hardly a trace of taste to it. It takes no purifying, and there is no problem of corrosion."

One of the true benefits of geothermal heating, I had discovered earlier that morning, is its absolutely pollution-free character. I had awakened at dawn and looked from my hotel room window across the city of Reykjavík. On a bitingly cold February day, when every structure in the city would have its heating system running at the fullest, not one plume of smoke, not one trace of smog hung in the air.

Curious, I asked Dr. Pálmason about the cost to the consumer of Iceland's geothermal energy — and discovered another benefit. "A homeowner might pay $20 a month on the average for heating bills. This is about one-fourth of the cost of heating a house with oil."

That afternoon, I joined Dr. Ingvar Birgir Fridleifsson, a young geologist with the energy authority, on a trip to the drilling site. Along the way, he explained why there is such an abundance of hot water in Iceland. "As you know, we are situated on the Mid-Atlantic Ridge, so there is a source of heat close at hand. Rainwater and snowmelt seep down tilted layers of permeable rock that overlie slabs of impermeable rock. As the water descends, it becomes heated; what we do is find it at the proper depth and pump it out." Ingvar showed me several pumping stations and the long concrete-covered pipes through which the water is channeled to Reykjavík. "The insulation is so good in these pipes that there is little heat loss during transport."

I marveled at the simplicity and effectiveness of the Reykjavík system. "We Icelanders take it as a point of pride to use our natural resources wisely," Ingvar said. "My father-in-law, for example, uses his hot water first to heat his house, then to warm a small, sheltered flower garden, then to heat a greenhouse, and finally for warm water in his swimming pool. Many people consider it almost a duty to try to use every calorie of heat from the water."

In their efforts to utilize all available sources of energy, the Icelanders of Vestmannaeyjar, the town struck by eruption in 1973, are even drawing heat from the lava that partially buried their town.

Sveinbjôrn Bjornsson of Iceland's Science Institute explained the process to me. "The lava flow is approximately 350 feet thick; its interior temperature reaches 1,800° F. Rainwater that falls on it quickly penetrates and is turned to steam. We're using the simplest of ventilation systems to suck this steam from the lava and use it to heat water. This water is then circulated into town. Already the new hospital and some 25 houses are heated this way, and the plans are set for converting the rest of the town. We estimate that we can use the lava flow to heat Vestmannaeyjar for 20 or 30 years."

As I listened to these plans, I reflected that the rest of the world could learn much about energy use and conservation from the Icelanders. Although they live in a country built by fire—by "mountains gone mad"—they regard that fire not as a threat, but as a resource yielded by one of the awesome powers of nature.

Fertile farmland of weathered lava produces sugarcane crops on Hawaii. Spinning off chaff, a "chopper-cleaner" harvests cane on rich and productive volcanic slopes. Ferns and lichens (below) establish first growth on a Hawaiian flow some 70 years old.

By Judith E. Rinard

The Weather Machine: Rain and Snow, Wind and Storm

"Blow, winds, and crack your
 cheeks! rage! blow!
You cataracts and hurricanoes,
 spout
Till you have drenched our steeples,
 drowned the cocks!"

—Shakespeare, *King Lear*

"I was sitting on the porch with four of my children," said Mrs. Georgia May, recalling the afternoon of April 4, 1977, in Birmingham, Alabama. "Suddenly, I saw this huge black cloud coming right at us. The wind began blowing, and the sky got blacker and blacker.

"We ran inside the house and crowded into a closet, just as the tornado hit. The whole house exploded. It was like a bomb going off. We fell down through the closet floor into the basement, and the house crashed in on us. Somehow the clothes in the closet cushioned our fall. When we managed to crawl out through a hole in the wreckage, all the houses had been blown away. It looked like the end of the world."

The storm Mrs. May was describing had occurred only a few days before. Her young face was still blank with shock and disbelief. Twenty-two of her neighbors had been killed. Hundreds more were left homeless.

The storm that hit Birmingham was a "maxi"—the largest, most lethal of all tornadoes. Masked by heavy thunderstorms, it struck at 2:55 p.m. Within 15 minutes, it plowed a path 15 miles long and nearly a mile wide through suburbs on the city's northwest side. I saw trees stripped of their leaves, their tops snapped off. Many had been torn out by their roots and hurled through the air into houses. Cars, trucks, and buses lay crushed like empty cans.

"That's typical F5 damage," Jim Campbell of Birmingham's Weather Office told me. "The F stands for the Fujita-Pearson scale, a rating classification of 0-to-5, based on intensity, wind speed, and the length and width of a storm's path. F5 tornadoes are any with winds of more than 260 miles an hour.

"Experts who surveyed the damage here tell us that this tornado probably reached wind speeds of about 300 miles an hour. Such winds, combined with the suction from that kind of storm, can pick up 80 tons.

"Usually," Jim continued, "a basement or small interior room like a bathroom will give some protection. But in an F5 wind there is no place to hide except underground."

As I walked through the rubble, I saw steel girders wrapped like ribbons around telephone poles; mattresses perched in tree branches; great wooden beams driven deep into the ground; chickens wandering about, their feathers blown out.

Most of the survivors had left their ruined homes to stay with relatives and friends. But in the heart of the wreckage, I found Maxine Dowdell and Hilliard Wiley standing outside a makeshift shack. Next to the shack was the flattened heap that had been their home. They told me that when they heard the tornado warning on the radio, they ran to the storm pit they had dug across the street from their house.

"As soon as we got in," Maxine said, "the tornado hit and the wind roared over us like freight trains. It was over in about a minute. When we came out, our house was gone."

Tornadoes are nature's most violent, destructive storms over a small area. They range from thin ropelike funnels a hundred feet wide to howling black maxis more than a mile across. Spawned by thunderstorms, tornadoes sometimes occur in families. A large parent thundercloud called a supercell may spin off several tornadoes in succession.

"Twisters," as tornadoes are often called, occur in many parts of the world, but are especially common in the U. S. Some 600 to 1,000 hit the nation each year, most in a broad belt called "Tornado Alley," which stretches from Texas to Michigan.

Although tornadoes usually last only minutes, they are awesomely destructive. Since 1916, when meteorologists began keeping records, tornadoes have killed more than 11,000 people in the United States and caused billions of dollars' worth of damage. The most devastating twister in history was the tri-state tornado of March 1925. It traveled 220 miles across Missouri, Indiana, and Illinois, killing nearly 700 people.

What sets the winds of a tornado in motion? Scientists are not really sure. But forecasters do know the kinds of weather conditions that are likely to produce the severe thunderstorms that generate tornadoes.

At the National Severe Storms Forecast Center in Kansas City, Missouri, tornadoes and other severe storms are forecast nationwide. The center is part of the U.S. National Weather Service, an important unit of NOAA, the National Oceanic and Atmospheric Administration of the Department of Commerce. There I talked with Fred Ostby, Deputy Director of the center.

Tornado forecasting is done the year around, Fred told me, because tornadoes can strike in any month. Although they are sometimes generated in the fall by hurricanes that hit land, most U.S. tornadoes occur in spring and summer.

During this period, the Northern Hemisphere gradually faces the sun's rays more directly and for a longer time each day. The sun soon heats surfaces that had been cooler during winter. Warm, moist air from the Gulf of Mexico begins to push rapidly northward where it meets much colder polar air masses moving south. When air masses with contrasting temperatures and humidities collide, thunderstorms and their violent offspring, tornadoes, occur.

In the main operations room of the Severe Storms Forecast Center, men looked for potential tornadoes, preparing and examining scores of charts showing all the information received by the center from radar, weather balloons, ground reports, and satellite photographs.

When warm, moist air is pushed violently upward in the atmosphere, a thundercloud rapidly develops. The movement of the warmer air to colder regions aloft sets off the powerful (Continued on page 97)

DREW LEVITON

"There wasn't too much wind right here, but boy, over there!" Alabaman George Glover indicates the path of the tornado that hit the Birmingham area on April 4, 1977. Twenty-two people died and hundreds lost their homes. Damage reached 15 million dollars. Glover recalls seeing parts of roofs and buildings hurtling through the air. Bent, twisted trees mark the path of the storm.

Preceding pages: Almost 800 feet wide at the base, a tornado sweeps through a Kansas field in August 1974. Scientists who studied movie film of the storm measured winds of 159 miles an hour.

Splintered houses lie alongside untouched homes in the wake of a tornado in the suburbs of Birmingham, Alabama. Skipping and twisting with explosive force, the storm lasted just 15 minutes on the afternoon of April 4, 1977. Its vicious winds slammed automobiles about until only twisted masses of metal remained. Debris and fallen trees blocked roads and highways, hampering rescue efforts. At right, Andra May sits disconsolately in the wreckage of his home. The chicken at far right, remarkably still alive, lost most of its tail and back feathers to the twister.

In the midst of disaster a new life begins. Hilliard Wiley feeds a colt born shortly after the Birmingham tragedy. Wiley promptly named the animal "Tornado." Bandages on the mother's legs cover injuries suffered during the storm. At bottom right, Maxine Dowdell shows how she took refuge in a shallow but effective storm shelter when she heard the tornado approach. Before the twister struck, their home stood among trees behind a wooden fence. After the storm, only jumbled litter remained. Tornado damage results not only from the extremely high winds but also from intense low pressure at the storm's vortex. When a tornado strikes a building, especially a closed one, the low atmospheric pressure outside creates an imbalance that makes the high-pressure air inside burst outward—the building explodes.

COURTESY OF HILLIARD WILEY AND MAXINE DOWDELL (ABOVE). DREW LEVITON

Tornado breeder: A swirling mass of clouds advances slowly over the plains of Kansas. Such storms often produce multiple tornadoes. Meteorologists do not fully understand what causes tornadoes. They

Inner Workings of a Killer Storm

A large tornado often harbors multiple suction vortices inside a column of dust. Such vortices measure less than 30 feet across but spin faster than the tornado itself. The diagram at far right shows how scientists determined wind speeds within the tornado pictured on page 86. They clocked the movement of patches of dust (arrows) on successive frames of movie film. Highest wind speed shown here: 71 meters per second, or 159 miles an hour.

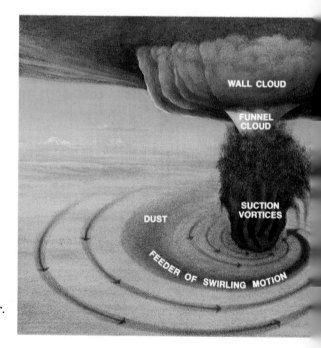

WALL CLOUD

FUNNEL CLOUD

SUCTION VORTICES

DUST

FEEDER OF SWIRLING MOTION

believe twisters form on the plains and in the Midwest when rapidly rising, ground-heated moist air collides with cool dry air. Some 600 to 1,000 tornadoes strike the United States each year.

APPARENT DUST MOTION *from Dundas movie*

updrafts and downdrafts in the thundercloud. These air currents release energy, sometimes forcefully, in heavy rain, lightning, hail, strong winds and, occasionally, tornadoes.

Such severe weather is the subject of intensive investigation at NOAA's National Severe Storms Laboratory in Norman, Oklahoma. Here scientists and researchers are working with experimental instruments called Doppler radars.

Unlike conventional radar, which detects only precipitation, a Doppler radar picks up shifts in wind velocity through echoes received from rain and hail. The wind shift indicates either a tornado or a thundercloud that may spawn a tornado.

Researcher Jean Lee told me that Doppler radars have identified tornado-spawning clouds more than 30 minutes before public sightings of the tornadoes themselves.

Dr. T. Theodore Fujita of the University of Chicago is widely recognized as one of the leading tornado experts in the world today. Born in Japan, Dr. Fujita became interested in tornadoes when he was a physics professor in his native country and read of a "dragonwhirl," a colloquialism in Japan for tornadoes. He came to the United States in 1953.

Today he heads the university's Tornado Research Project. His work has led to many discoveries and earned him the nickname "Mr. Tornado." When I mentioned the name, he laughed and said, "I hope people who hear me called that don't think I bring tornadoes."

Dr. Fujita's research ranges from laboratory experiments with an ingenious "tornado machine" of his own invention to studying the effects of real tornadoes. Like a meteorological Sherlock Holmes, he has formulated investigative methods to uncover the secrets of his quarry. "Tornadoes," he says, "are like criminals who cannot get away without leaving their fingerprints."

Every year, Dr. Fujita and his staff survey the damage of tornadoes and take a series of photographs both from the air and on the ground. Using this method, Dr. Fujita discovered

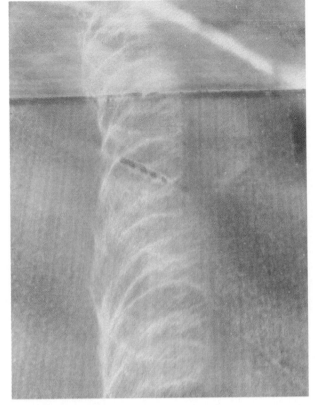

Whirlwind in a laboratory: At the University of Chicago, Professor T. Theodore Fujita studies a Lilliputian twister. He designed the storm machine, which stirs up dry-ice vapor to simulate tornadic action. According to Dr. Fujita, multiple vortices in tornadoes account for the skipping effect that can bypass one house and destroy a neighboring one. The spiraling lines of a tornado's ground track (above) support his theory. In 30 years of studying tornadoes, Fujita has never seen one—thus his "scoreless" license plate.

DONALD J. CRUMP, NATIONAL GEOGRAPHIC STAFF (BELOW). NOAA, JOSEPH H. GOLDEN

the existence of smaller twisting funnels within large tornadoes. These small twisters, or "suction vortices," leave telltale spiraling lines on the ground that correspond with the worst damage in the course of the main tornado. "It is these suction vortices," he said, "that have the strongest wind speeds — up to 300 miles an hour. They are what pick up automobiles and do F5 damage."

The Birmingham tornado had such extreme winds. "It was one of the strongest tornadoes I have ever studied," Dr. Fujita told me.

In his laboratory, Dr. Fujita demonstrated his tornado machine. Beneath a large dome, several layers of metal cups spin around and around, slowly on the outside and faster in the middle, to simulate the suction and winds of a rotating thundercloud. Below sits a pan of water.

As I watched, Dr. Fujita threw several pieces of dry ice into the water. When the water began to froth and bubble, he switched on the "wind power." A little white funnel sprang from the pan and twisted upward — a miniature tornado.

Next, Dr. Fujita tossed several chunks of Styrofoam into the water. Almost instantly, the little twister sucked the pieces up and spat them violently out again, clear across the room. When Dr. Fujita placed a grillwork of heated wires across the pan of water, the funnel became disorganized. Soon it broke up and dissipated entirely. Then he reactivated the twister and placed a ruler vertically inside it. This also caused the funnel to break up.

Dr. Fujita pointed out that such experiments may suggest possible methods of tornado modification. However, he cautions, much more research is needed. "We are not ready to modify real tornadoes now, but yes, I believe one day it will be possible."

While researchers like Dr. Fujita are probing the mysteries of tornadoes, one scientist has turned to the sea in search of clues. I talked with Dr. Joseph Golden at NOAA's Environmental Research Laboratory Headquarters in Boulder, Colorado. Since 1965, Dr. Golden has been

Bulbous cloud protrusions often accompany thunderstorm systems that produce tornadoes. These clouds appeared in Oklahoma, a state often hit by tornadoes. A waterspout — essentially a tornado at sea — leaves a spiral pattern on Florida Bay (opposite).

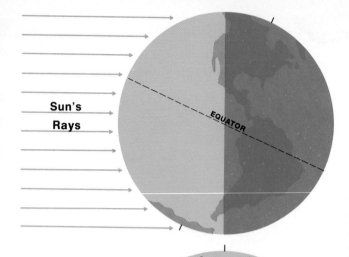

**Sun's
Rays**

EQUATOR

EQUATOR

What Makes
Our Weather?

*Symbols superimposed on a
satellite photograph of the earth
show weather conditions for
September 1, 1977. As the jet
stream—a belt of high-altitude
winds—loops eastward over
North America, it mixes cold air
from the north with warmer air
from the south, causing large
eddies called high and low
pressure systems. These highs (H)*

Energy from the sun powers the
"weather machine." It heats our
spinning earth and churns its
atmosphere, causing our winds
and weather. The sun also
produces the seasons, as it strikes
the tilted, orbiting earth. Air near
the Equator, rapidly heated, rises
and flows toward the Poles, while
colder polar air sinks toward the
Equator. These winds would flow
directly north and south if the
earth remained stationary (broken
arrows). But its easterly rotation
deflects and curves the winds
(solid arrows). They move gigantic
air masses that collide and mix,
producing our changing weather.

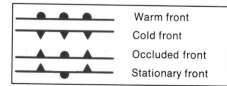

	Warm front
	Cold front
	Occluded front
	Stationary front

—Jet Strea

L

No

and lows (L), moving west to east with the winds, carry storms and fair weather across the continent. Arrows indicate the direction of the winds on the day of the photograph. In the Northern Hemisphere, winds whirl clockwise around highs and counterclockwise around lows. In the Southern Hemisphere, wind directions reverse. Moisture-laden trade winds collide over the oceans near the Equator, creating bands of storminess and clouds. At fronts, where air masses of different temperatures and humidities meet, changing weather occurs. Cold fronts indicate a cold air mass advancing on a warmer air mass. Warm fronts occur where warm air replaces retreating cold air. Stationary fronts settle where two air masses move very slightly or not at all. An occluded front occurs where a cold front overtakes a mass of warmer air and pushes it upward. White areas on the photograph indicate clouds. In the Gulf of Mexico Hurricane Anita, the first hurricane of 1977, spins toward the Mexican coast. Born in a low pressure disturbance off Africa, it transports heat away from the warm tropical ocean.

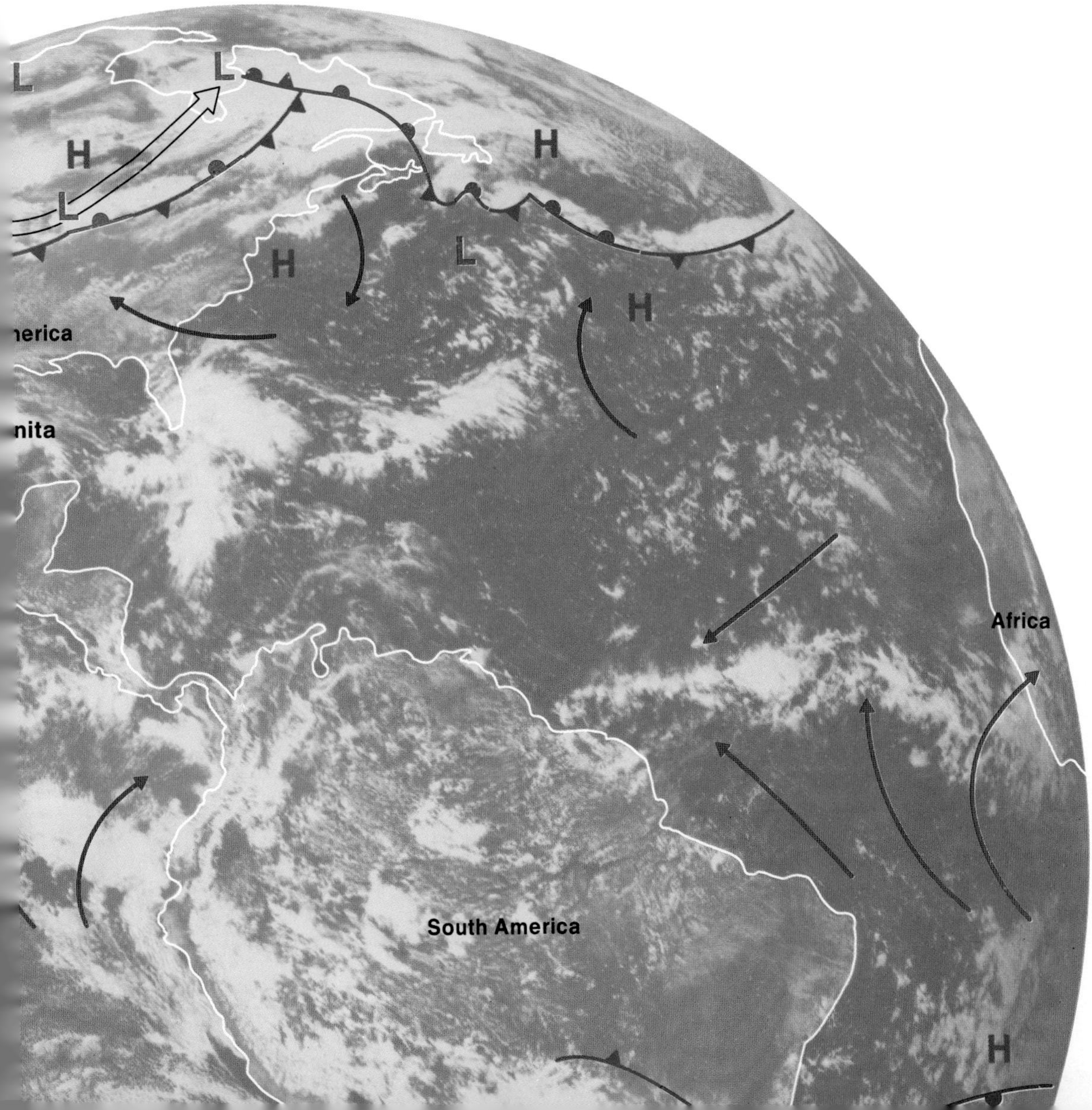

DIAGRAMS BY PETER J. BALCH AND MARGARET A. DEANE, BASED ON DATA FROM NOAA

studying waterspouts, cousins of tornadoes, that occur over water.

Dr. Golden believes that waterspouts and tornadoes have many similarities, and that waterspouts are likely testing grounds for attempts at tornado modification.

"Attempting to modify tornadoes on land would be very dangerous, I think," he said. "Tornadoes and the conditions that produce them are extremely violent." But waterspouts, he pointed out, are usually less violent and occur well away from land.

Tornadoes are only one of the forces powered by the mysterious and intricate machine we call weather. Hailstorms, blizzards, and hurricanes also rake the earth. Like all forms of weather, storms have a profound effect on human lives and property. They serve a necessary and beneficial function in the overall weather pattern, but sometimes are enormously destructive.

In simplest terms, weather is a giant mechanism for distributing the sun's heat. At the Equator, the earth receives more energy from the sun than it loses into the atmosphere. At the Poles, more energy dissipates into the atmosphere than is received from the sun. If it weren't for the winds, the tropics would be unbearably hot and the polar regions unbearably cold. The winds that blow in great revolving belts around the earth are like conveyors that carry excess heat from the Equator toward the Poles and cold polar air back down toward the tropics.

Most of the air does not flow a nonstop route. Warm air traveling toward the Poles goes only about a third of the way; then it cools and sinks and travels back toward the tropics as the cool trade winds. Similarly, most of the polar air moving toward the Equator grows warmer a third of the way down and begins to rise and flow back toward the Poles.

Between these two belts of air is another one over the mid-latitudes of the earth. The planet's rotation, which reaches a speed of about 1,100 mph at the Equator, makes this wind blow toward the east. The earth's spinning also deflects the winds flowing from and to the Poles. Winds circulating between the tropics and the Poles are called westerlies at low altitudes, the jet stream at high altitudes, and trade winds if they travel back toward the Equator.

In the Northern Hemisphere, the fast-moving belt of wind called the jet stream flows from west to east five to ten miles above the earth, traveling in a looping pattern, like a great waving ribbon. As the stream travels around the earth, it causes great eddies called high pressure systems that rotate clockwise between its looping waves. Between the "highs" are low pressure systems that spin in the opposite direction. As these highs and lows move across the middle latitudes with the westerly winds, they bring alternating systems of fair weather and storms. Storms occur along lines called "fronts" where air masses with different temperatures and humidities collide.

Every day, the sun's heat evaporates trillions of tons of water from oceans, lakes, and rivers. The tremendous energy required to evaporate so much water is stored in the air. When the water vapor is cooled, it condenses, and clouds form. They may be blown for miles before releasing their energy in rain, snow, or the lightning of a thunderstorm.

Some 1,800 thunderstorms are taking place over the earth at any given time. Energy equal to a small hydrogen bomb can be generated during the course of a thunderstorm lasting several hours. A hurricane can release almost that much energy every second.

Distributing the sun's heat over the earth is the principal function of the weather machine. Storms also transport water from place to place.

A spectacular by-product of thunderstorms is lightning. "Most people don't realize it, but lightning kills more people in the United States every year than tornadoes or hurricanes," said Edwin Weigel, Public Affairs Spokesman of the National Weather Service, near Washington, D. C. "Lightning is responsible for about 150 deaths in the U. S. annually, and about 300 injuries." Lightning also kindles 10,000 to 15,000 forest fires a year throughout the nation, most of them in the Rocky Mountains and Alaska, and on the Pacific Coast.

In June of 1977, a stunning assertion of the powers of nature occurred in New York City. Lightning bolts struck several power lines in little more than an hour, knocking out the electrical system and plunging the city into darkness.

At NOAA's Atmospheric Physics and Chemistry Laboratory in Boulder, I asked Dr. Heinz Kasemir about efforts to modify the effects of lightning. "Basically," he told me, "lightning is the result of a huge buildup of electrical charges in a thunderstorm. There are always a nearly equal number of positive and negative charges—called ions—in the atmosphere. Many scientists think that negative ions are carried to the bottom part of a cloud by precipitation; positive ions remain at the top. Air is a good insulator and prevents the positive and negative charges from flowing together and neutralizing each other. So an electrical field builds up in the cloud.

"When the difference in voltage between the top and bottom of the cloud gets too high, the resistance of the air breaks down, and a conduction path forms between the two charge areas. This produces lightning. It may travel within the cloud from top to bottom and back, or from one cloud to another, or from cloud to ground and back again."

When lightning is touched off, the air becomes suddenly heated, and it expands with an explosive sound—thunder. A lightning bolt may discharge as many as 100 million volts of electricity.

Dr. Kasemir told me that the main reasons for lightning suppression experiments are to prevent forest fires and to safeguard spacecraft.

"In 1969," he said, "the Apollo 12 capsule was struck by lightning just after launch. Now every time a launch is scheduled, specialists assess the possibility of lightning and *(Continued on page 110)*

Not quite fogged in, an airliner takes off from National Airport at Washington, D. C. Weather can play havoc with schedules, so accurate forecasts play a vital role in airport operations. At the National Center for Atmospheric Research in Boulder, Colorado, scientists program a computer to help develop more reliable short-range forecasts.

Following pages: Lightning bolts streak the sky above Tampa, Florida, in this multiple exposure. Lightning can pack 100 million volts and temperatures five times hotter than the sun's surface.

Human lightning rod: Retired park ranger Roy C. Sullivan, hit by lightning seven times in 35 years, indicates where a bolt struck him in the shoulder in 1970. Lightning has severely burned Sullivan and damaged his hearing. In a rare occurrence at a lightning research tower near Lugano, Switzerland, a bolt leaps skyward (lower left). Another bolt strikes a tree nearby. Burn lines streak a golf green in Arizona where lightning burst after hitting a fiberglass pole.

PHIL KRIDER (BELOW). RICHARD E. ORVILLE

PHIL SCHERMEISTER (OPPOSITE). FOLLOWING PAGES: NELSON MEDINA, PHOTO RESEARCHERS

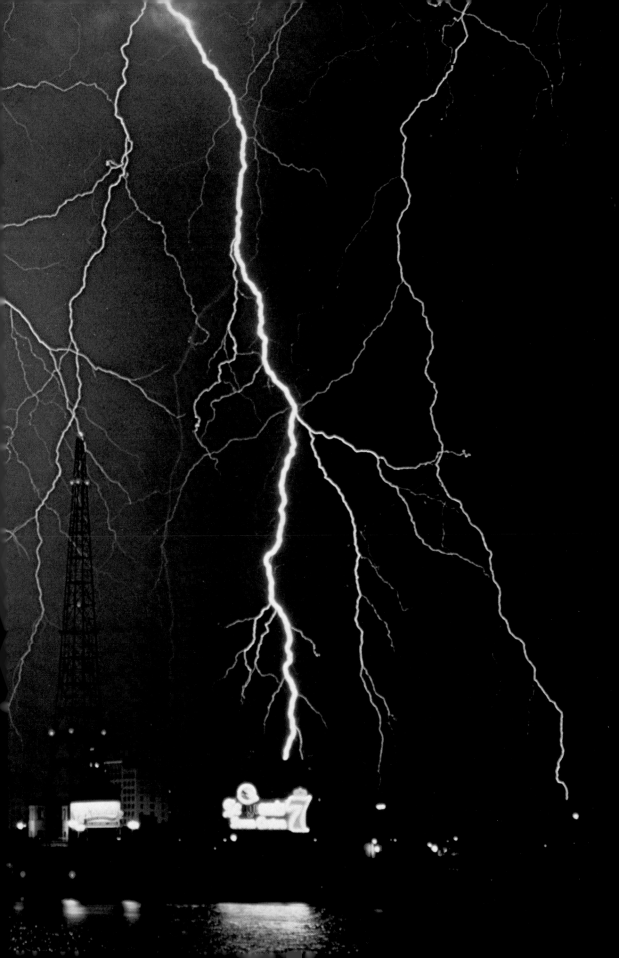

Hail plummets from a thunder-
head on the plains of eastern
Colorado. Hailstones, shown in
cross section under polarized light
(opposite), reveal how hail forms:
Rings grow around a central
core—usually an ice particle—that
travels through varying
temperature zones within the
storm cloud. The particle, carried
by strong updrafts and
downdrafts, alternately rises and
falls. The rings reflect the growth
in different parts of the cloud.
Children (opposite, upper) hold
hailstones from a storm that pelted
Coffeyville, Kansas, in September
1970. The largest hailstone ever
recorded fell during the same
storm; it weighed 1.67 pounds and
measured $17^{1}/_{4}$ inches in
circumference.

BOB STONE (ABOVE). HAILSTONES: CHARLES KNIGHT, NCAR   JAMES A. BOONE

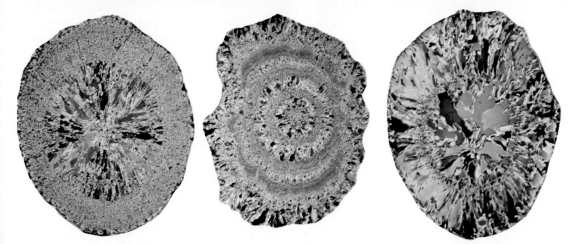

tell space officials whether it is likely a strike could occur."

In his laboratory, Dr. Kasemir showed me a mechanism used in experiments in lightning suppression. When he turned on the machine, sparks began to fly in a fiery line between two spheres—one representing the earth and the other a cloud—a miniature simulation of lightning. He picked up a plastic rod with a piece of metal fiber as thin as spider's web attached to it. When he held the metal fiber between the two spheres, the sparks immediately stopped. When he took it away, the sparks flew again. "Invisible ions from the metal fibers absorbed the charge as it flew between the two spheres," he said. "That shows the principle of our method in a nutshell. By seeding the bottom of thunderclouds with millions of these tiny fibers, it's possible to provide many small conductors so the electrical charges dissipate before lightning can occur."

Although lightning most frequently hits high buildings and towers, it sometimes strikes human beings. On a steamy day in June, I paid a visit to Roy C. Sullivan at his home in Waynesboro, Virginia. A stocky, white-haired man, Roy welcomed me to his home just outside the Shenandoah National Park, where he served as a ranger until his retirement in 1976. He looked remarkably robust for a person who had been struck by lightning six times.

Roy holds the distinction of being listed in the *Guinness Book of World Records* as the champion survivor of lightning. His closest rivals were people who had been struck as many as three times. In every case, the third strike was followed by a funeral.

"Some people are allergic to flowers," Roy told me. "I guess I'm just allergic to lightning." His neighbors call him "Lightning Rod."

It was in April of 1942 that Roy was hit by lightning the first time. He was at work in a fire tower when a thunderstorm closed in. "Lightning hit the tower seven or eight times," he told me, "and fire was bouncing all around, so I decided to bail out. I had gone just a few feet from the tower when it struck me. It went down my right side and burned a half-inch strip all the way down my leg and knocked my big toenail off. It burned a hole through a pocket watch I had in my pocket. I still have scars on my

ANNIE GRIFFITHS

Minnesota farmer looks over his crop after a storm that brought hailstones as big around as half dollars. Hailstorms cause 600 million dollars in crop damage yearly in the United States.

110

right arm and leg from that strike."

Another strike occurred when he was on duty at the Park Registration Station. "There was a gentle rain," he said, "but no thunder until suddenly one big clap—the loudest thing I've ever heard. Fire was bouncing around the station, and when my ears stopped ringing I heard a sizzling noise. It was my hair—on fire."

"Before the lightning strikes, do you have any warning at all?" I asked. "Well, you can tell, but it's too late. You can smell sulfur in the air, and then your hair will stand up on end, and then it's going to get you. You don't have time to do anything."

A few weeks later the phone rang in my office. It was Roy calling me from the hospital three days after his *seventh* lightning strike. "I'm feeling better now," he said, "but it was pretty bad. It was a hot one. I was out fishing when it happened. I smelled sulfur and looked up and saw the bolt coming down to get me. Blam! I hope this is the last time. Seven times is enough. In fact, it's too many."

Although not so dramatic as lightning, hail is one of the most destructive of all weather phenomena. Each year, it destroys some 600 million dollars' worth of crops in the United States. For many years, scientists around the world have been working to control it.

Dr. Charles Knight, a hail expert at the National Center for Atmospheric Research (NCAR) in Boulder, is one of the leaders of the National Hail Research Experiment, a project involving a number of government agencies and universities.

In the 1960's, Dr. Knight explained, the Soviet Union reported a 90 percent success rate in suppressing hail by cloud seeding. This report stimulated intensive hail research in the United States, as well as in several other countries. The fieldwork is done in "Hail Alley," the region in the U. S. with the highest frequency of hailstorms. It is centered on the point where Colorado, Nebraska, and Wyoming join.

In 1972 the Colorado research group began seeding hailstorms, using a technique based on the Soviet theory that hail grows primarily from supercooled droplets in thunderstorms. By seeding the bases of the clouds with tiny particles of silver iodide, the theory holds, more nuclei will compete for the moisture and prevent the droplets from growing into large hailstones. Instead, rain or sleet will result.

Although this theory apparently worked well in the Soviet Union, the U. S. researchers found that it did not seem to apply to the storms in Hail Alley. "We found by analyzing the hailstones that 80 percent of the hail grows around graupel, or ice-covered snow crystals, instead of frozen raindrops," Dr. Knight said. "Apparently this type of hail is more difficult to suppress with cloud seeding."

A further complicating factor was the discovery that hailstorms themselves are not uniform. "There are several different types of storms that produce hail," Dr. Knight pointed out, "and quite possibly each type responds differently to seeding."

The more I learned about the complexity of weather, the more I came to appreciate what a difficult and intricate job forecasting is. All manner of weather phenomena must be considered in forecasting—and on a global basis. For the weather that affects North America has its origins in distant oceans and lands and atmospheric currents.

Benjamin Franklin was one of the first in this country to observe that weather moves across the land in predictable patterns. But it was not until Samuel Morse's invention of the telegraph in the mid-19th century that information about approaching weather conditions could be communicated across country before the weather itself arrived.

The National Weather Service, formerly the Weather Bureau, began making forecasts for the U. S. in 1870. By then, meteorologists realized that the key to predicting weather lay in gathering data about the atmosphere as it moved over the earth, and analyzing it to project how it would affect future weather.

At first, weather observations were relayed by telegraph from

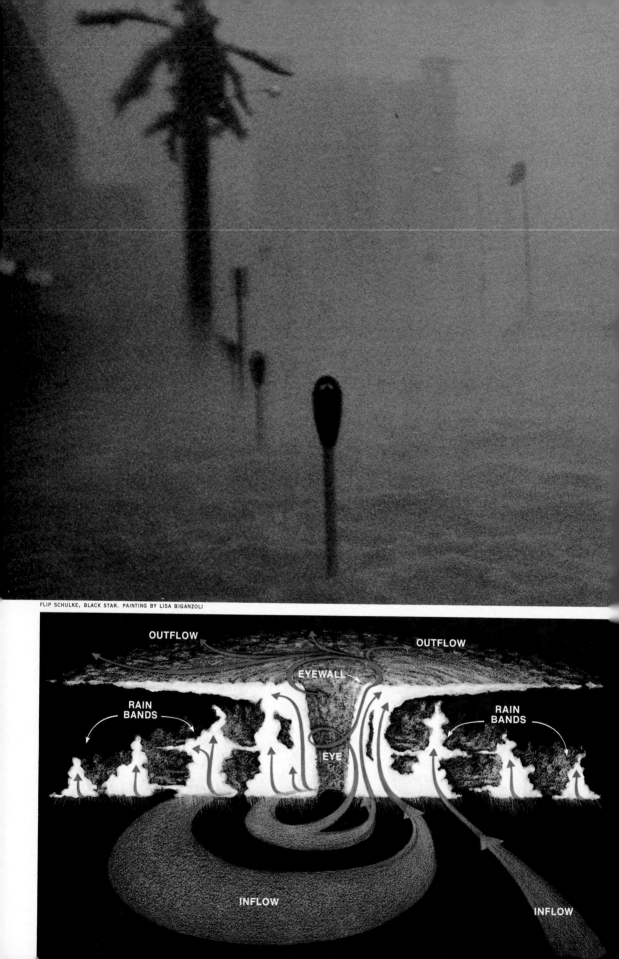

FLIP SCHULKE, BLACK STAR. PAINTING BY LISA BIGANZOLI

OUTFLOW

OUTFLOW

EYEWALL

RAIN
BANDS

RAIN
BANDS

EYE

INFLOW

INFLOW

Hurricanes: Whirling Dervishes Out of the Tropics

Savage winds of Hurricane Betsy drive ocean water into downtown Miami. During the 1965 storm, more than 70 people lost their lives in Florida and along the Gulf Coast. Though tornadoes have stronger winds, hurricanes carry a greater threat to life and property because of the storm surges— monstrous wind-driven tides— and torrential rainfall they cause.

Meteorologists describe hurricanes as giant heat engines. In the Northern Hemisphere, hurricane winds flow counterclockwise (painting) and with increasing velocity converge on the "eye." There, moisture-laden winds spiral upward; as the air cools, it releases rain and enough heat energy to supply the U. S. with electricity for six months. Around the center, a wall of clouds rises thousands of feet. Here, inside the eye, winds seldom reach speeds higher than 15 miles an hour. Rain bands, with thunder and lightning, form in spirals around the storm.

Like swirling dancers, two Pacific hurricanes spin around each other—a phenomenon known as the Fujiwhara effect—as they move slowly off the coast of Mexico. A satellite photograph shows Ione, at lower left, and Kirsten in 1974.

around the country to the headquarters in Washington, D. C. But the data were so incomplete, and the knowledge of the atmosphere's behavior so limited, that early forecasts were little more than guesses. Even today, many people joke about the weatherman's fallibilities, and one mistaken forecast is likely to be remembered while a long record of accurate forecasts is forgotten.

Actually, today's short-range weather predictions are amazingly accurate. As Ed Weigel told me, "Forecasts are about 90 percent accurate for one day ahead and 80 percent accurate for two days."

Each day, the National Weather Service collects data from more than 1,000 stations across the country. Reports on temperature, air pressure, relative humidity, and wind speed and direction come both from stations on the surface, and from a few in the upper air. Thousands of other reports come from aircraft, ships, buoys, and foreign stations. Satellites orbiting the Poles and others poised over the Equator take pictures of the earth's cloud cover by day and by night.

All of this data is fed into computers that analyze weather conditions, according to Dr. Warren Washington, a research scientist at NCAR. By using fast computers so that a forecast runs ahead of real time, meteorologists get predictions of what the conditions are likely to produce as the atmosphere carries its weather systems across the globe. The computers' projections, coupled with the skill of meteorologists who know the patterns of past weather and the special conditions that can influence moving air masses, produce the final forecast.

But these methods are reasonably reliable for only two days, since basically the information collected and analyzed covers only the Northern Hemisphere. The longer the forecast period, the more the forecaster must know about conditions all around the globe. Today, according to Dr. Washington, a five-day forecast is much less accurate than a two-day forecast, while a 30-day forecast is "almost a gambler's hunch."

For months I had been traveling around the country, learning a great deal about the complexity, mystery, and power of the weather machine. At least I thought I had, until one dark morning in September, when I found myself flying straight into the core of one of nature's most terrifying forces—a hurricane.

On September 1, 1977, Hurricane Anita was whirling wildly over the Gulf of Mexico. Reports indicated that it was growing into the strongest storm in years, possibly the worst since Camille in 1969. If so, she could be a killer. Finding Anita and tracking her path demanded urgency.

Shortly after midnight of September 2, I boarded a huge WC-130 aircraft, aptly named the Hercules and now used by the Air Force to track and penetrate hurricanes. The planes operate out of Keesler Air Force Base in Biloxi, Mississippi.

When a developing hurricane is spotted by satellite, the National Hurricane Center in Miami sends out a reconnaissance aircraft to observe the storm, using satellite photographs for reference. The next stage is to penetrate, measure, and track the progress of the hurricane.

The crew was part of the 920th Weather Reconnaissance Group, known as the "Storm Trackers." They are a volunteer unit of Air Force Reservists, and they fly 70 percent of all hurricane reconnaissance missions. The planes penetrate the calm eye of the hurricane at about 10,000 feet. Once inside, the crew releases a parachute-borne metal cylinder called a dropsonde. It contains instruments to measure the air pressure, temperature, and humidity inside the storm. A radio transmitter in the cylinder sends the data back to the aircraft.

As I sat in the plane waiting for takeoff I thought about what Col. Chuck Coleman had told me during my briefing an hour earlier. "The WC-130 is a very strong aircraft; we haven't lost one in the two years we've been flying, but other recon groups have lost three since 1947. So there is a chance it could go down. If you do have to ditch in the ocean, I must tell you that you probably won't need your lifejacket—the chance of survival is virtually nil."

At 1 a.m. the big plane lumbered down the runway and took off into the dark sky. I had been given earphones so I could listen to the conversations of the pilot and navigator. For a while, I could hear only the deafening pounding of the four engines, and see nothing out of the window behind the cockpit but total darkness. Eventually, however, I began to see ominous flashes of blue lightning over the black water. On my headset I could hear the aircraft commander, Captain Tom Mish, telling the crew that it was time to strap in and to tie down any loose objects. It was 2:45 a.m.

A crew member told me what to expect. The hurricane's strongest winds, he said, are just around the eye, especially in the right front quadrant. The plane would try to penetrate the weakest part of that wall, the left rear quadrant.

Minutes after the pilot's order to strap in, the heavy plane began to lurch and rock. Lightning lashed out in the black sky, and rain drove furiously past my window. Occasionally, the lightning lit up the ocean below us with an evil, flickering light.

The voice of the navigator crackled over the headset. "Looks like it's ahead of us, but the rain's so heavy I can't see anything on the radar. We're going to have to go through this rough stuff; get down there and get that computer working."

To find the hurricane's eye, the navigator uses a computer on the aircraft that continuously receives coordinates of the storm's location. Both Doppler radar and precipitation-scanning radar are on board to help locate the eye. But heavy rain kept weakening the Doppler signals.

At times it seemed as if engines, radar, and computer were all breaking down at once. I heard one exasperated voice after another from the cockpit. "We just missed it. Let's go back and try her again." "Let's sneak back around and try a left turn about ten degrees." "I'd almost think that eye doesn't exist. I think it's closed in." Then there was a flash of lightning, and the plane jolted up and

September 25, 1975

September 26

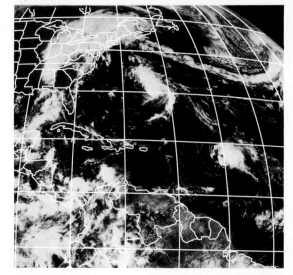

September 29

September 30

Tracked from a satellite 22,300 miles high, two powerful storms move across the western Atlantic. The hurricanes, Faye and Gladys, appear as tight spirals in this series of photographs. Taken on consecutive days, the pictures recorded the progress of the hurricanes as they moved northward over the Atlantic, off the U. S. coast. Neither of the hurricanes touched land.

down. "The thing's trying to throw us out," said the pilot. Heavy rain was pounding directly against my window now, obscuring it to a blur.

For more than an hour the big aircraft made repeated efforts to enter the hurricane. It was almost four in the morning when a hopeful note finally came over the headset. "Wait a minute, we've got something here up ahead that looks like a hole." Then I heard the pilot again. "Everybody tied down back there?" "Roger, everything's snug as a bug in a rug." "Good, I have a feeling it's going to be a little bumpy going in." Suddenly, the plane lurched up and down again as if it would surely fly apart. I sat in suspense as the pilot guided the aircraft

116

September 27

September 28

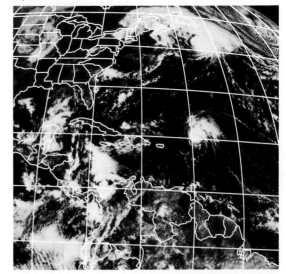

October 1

October 2

into the hurricane. The thrashings of the plane were more violent, and the rain was now a solid mass of water. I held my breath. Then, suddenly, the bumping stopped.

"We've got it. We're in the eye now. Looks like it's about 14 miles across." The pilot called me up to the cockpit where I could get a good view of the hurricane's eye.

We were in a swirling bowl of white, cottony clouds. It was like a huge stadium at night. In the pitch black sky above us, I could see the crystalline stars and, shining above all, the white disk of the waning moon. The plane droned along peacefully in the calm, strangely beautiful sanctuary of the hurricane's eye. It was as though we had cut through a dimension in time and space. For a magic interval, there was no fear, no hurry, no frustration—only a majestic calm.

We stayed in the eye for 45 minutes, long enough to fix the storm's position and relay data measured by the dropsonde to the Hurricane Center, and then we turned back. It was almost nine a.m. when we touched down at Keesler, after another rough ride, and I learned that, mercifully, Anita had not been a killer. Villagers at the little Mexican coastal town of La Pesca, where the storm raged inland, had been safely evacuated.

Hurricanes breed over tropical oceans during the warmest months of

117

"Hurricane-watch" party ends in tragedy. The Richelieu apartment building stood at Pass Christian, Mississippi, before the onslaught of Hurricane Camille in 1969. After the storm, only foundations and a swimming pool remained. Of the 25 partygoers, all but one— Mrs. Mary Ann Gerlach (opposite)—died in the hurricane. Other apartment dwellers fled before the winds struck.

the year and occur in many parts of the world. In the western part of the North Pacific, they are called typhoons; in the Southern Hemisphere and Indian Ocean, cyclones; and in the eastern North Pacific, Caribbean, and Atlantic, hurricanes. Whenever they slam into land, they bring a triple threat from lashing winds, torrential rains that cause rapid flooding, and monstrous tides of wind-driven water called storm surges.

Throughout history, hurricanes have caused more deaths than even earthquakes and volcanoes. In 1737, a cyclone hit Calcutta and killed 300,000 people. In 1881, the Haiphong typhoon swept over the low coast of China, also taking a toll of 300,000 lives. One of the greatest disasters of this century struck on November 12, 1970, when a cyclone roared over the coastal islands of Bangladesh, then known as East Pakistan. That storm killed at least 500,000.

In the United States, the average annual death toll from hurricanes and tropical storms has been about 190. In 1972 damage exceeded two billion dollars when Hurricane Agnes caused record floods in the northeastern United States.

Hurricanes are whirlwinds that may grow to be more than 300 miles in diameter. As they spin over the ocean, converting heat from the sun-warmed water to wind, hurricanes release more energy each minute than the largest hydrogen bomb.

Centuries ago, Caribbean Indians believed that Hurracan—God of All Evil—was responsible for the terrifying storms. If the god grew angry, he would send the "evil wind" and blow the people out like candles.

In the Northern Hemisphere, the hurricane season runs from June 1 to November 30. During those months the sun warms the tropical ocean water to a temperature of 80 degrees or more. Once the whirlwinds start moving, they feed on energy from the warm, moist air over which they travel. As this air spirals toward the eye of the storm it rises rapidly. At higher altitudes, the hot air cools and condenses, forming

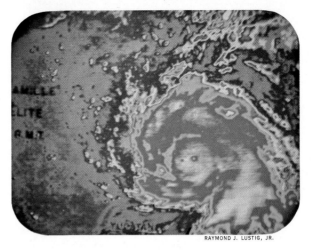

Portrait of a mass killer: Satellite photograph shows Hurricane Camille, a storm that claimed some 320 lives. Added color indicates the hurricane's moisture content; white marks areas of heaviest rainfall. Mrs. Mary Ann Gerlach, below, only survivor of the Richelieu apartment building hurricane party, points out the building's foundations. Storm waves washed Mrs. Gerlach from the wrecked building and carried her 4½ miles. The wind, clocked at 200 miles an hour, "was so strong you could hardly breathe," said Mrs. Gerlach.

swirling bands of clouds and rain and releasing heat energy. This energy fuels the hurricane's winds and drives them in a tightening coil around the eye at speeds of more than 200 miles an hour.

As long as a hurricane is over warm water, it usually continues to build. While moving over the ocean, it lifts billions of tons of water every day and dumps most of it back as rain in the next 24 hours. When the storm moves inland or over colder water, the loss of heat energy from the ocean eventually causes it to die.

In spite of the frightening devastation they bring, hurricanes are a positive force in nature, I learned from meteorologist Dr. Neil Frank, Director of the National Hurricane Center at Miami. These great storms, Dr. Frank told me, distribute excess heat and moisture from the tropics to the Poles.

Scientists believe that most Atlantic hurricanes begin as disturbances called easterly waves — troughs of low pressure that form in the trade winds moving west from Africa. About a hundred of these occur in the Atlantic each year. Only about ten of them grow into tropical storms, or potential hurricanes. When a tropical storm develops whirling winds of 75 miles an hour or more, it qualifies as a hurricane. Only two or three tropical storms and hurricanes reach the North American continent each year; the others weaken over the ocean and die.

A network of radar installations along the Eastern Seaboard from Maine to Texas monitors hurricanes when they approach within 250 miles of the coast. If a storm gets within 200 miles, aircraft penetrations are made as often as every three hours.

The worst natural disaster in U.S. history, Dr. Frank recounted, was the hurricane that hit Galveston, Texas, in 1900, killing 6,000 people. Since then, he added, loss of life from hurricanes has steadily decreased because of better warning and better communications systems. But damage from hurricanes has steadily gone up due to a great increase in building and development along beachfront property.

The area of greatest devastation, according to Dr. Frank, is seldom more than 50 miles wide. The intense winds around the eye push water inland in the storm surge, a dome of water that pounds the coast where the eye of the storm touches land. Nine out of ten people killed by hurricanes drown in the surge.

In 1969, Hurricane Camille smashed ashore at Bay St. Louis, Mississippi, with winds of 200 miles an hour. At Pass Christian, it built up a 25-foot-high storm surge. The last eight feet of water came up in less than 30 minutes. Most of the people in the area heeded the warning and got out. But some decided to ignore the danger and more than 320 lost their lives. In one instance, Dr. Frank told me, 25 people in an apartment building in Pass Christian, just across the highway from the beach, decided to have a "hurricane-watch" party and wait out the storm. Of those 25, only one survived.

"If you want to know what a bad hurricane is like," Dr. Frank told me, "talk to Mary Ann Gerlach."

I found her in Gulfport, where she still lives, although for a few months after Camille she moved away from the area because of the terrible memories of the storm.

The night of August 17, when Camille hit the Gulfport area, Mary Ann and her husband, Fritz, were planning to attend the hurricane-watch party. "We had been in Florida during hurricanes," she told me, "and we were looking forward to the party. When a hurricane comes, everyone gets off work, and you cook a lot of food and get together to wait out the storm. It's — you know — party time.

Shrimpboats litter a wharf at Aransas Pass, Texas, piled there in 1970 by the roaring winds of Hurricane Celia. Costliest hurricane ever to strike the state, Celia logged winds of 130 miles an hour, with gusts to 161. Damage exceeded $450,000,000.

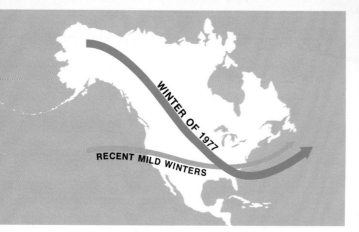

Ohio schoolchildren keep their coats on during the bitter winter of 1977. Dr. Jerome Namias (right) of the Scripps Institution of Oceanography believes the harsh weather resulted from a shift in the jet stream (diagram). Looping into Alaska, then far south, it caused drought in the West and cold in the East.

CHARLES O'REAR. DIAGRAM BY PETER J. BALCH

HERRAL LONG

Well, my husband and I both worked at night, and when we got off we were tired. So we decided to rest awhile before the party.

"When we lay down in our second-floor apartment, the water had come up to the swimming pool outside, but it was rising so gradually we didn't worry about it.

"Then, about 9:30 that night, I woke up and thought someone had broken into the apartment downstairs, because I could hear something moving around. By then, the water had come into the first floor of the building and was rising. It wasn't long afterward that we heard the wind and waves pounding at the picture windows in our living room.

"We went back into our bedroom, and in a few minutes we heard an awful popping sound as the windows broke. We held our shoulders to the bedroom door to try to keep the water from coming in, but it just came around to the back windows. In about five minutes the bed was floating halfway to the ceiling. I just knew then that I was going to die."

Mary Ann swam out of the window of the bedroom clutching a sofa pillow. She was swept into telephone wires outside and became tangled in them. She had a flashlight and flashed it on the apartment building. She saw Fritz go under the water. He never came up.

The building began to sway and crumble in the water. "I never thought a three-story building could sway like that," she said. It came crashing down, and Mary Ann kicked her way free from the wires to get away before the crumbling building could pull her under the water.

"Pretty soon," she said, "there was so much water I thought I had been swept out into the Gulf. Big trees and debris came sweeping over me. The wind was so strong that whatever I grabbed would just be ripped out of my arms.

"One time I almost didn't make it when a big tree came at me, loaded with debris. It pushed me under, and when I came up, the surface was covered solid with floating boards. So I had to dive deeper, and when I tried to come up again, the surface was still closed off with debris. Finally I managed to pry a space open with my legs and get to the surface again. That happened several times.

"I went over houses, trees, light poles—I mean there were two-story houses I was floating over. All around there was nothing but water."

Finally, she was swept into a treetop as the water began to go down in the morning. "I was about 12 feet up in a sort of nest of boards in the treetop," she said. "It was so cold, and the wind was still blowing hard."

At daybreak she passed out for a while, and woke up later when she heard people yelling. "The first thing I remember seeing was a human skull lodged in the tree branches," she told me. "It had been washed up from a cemetery.

"Then a man found me and carried me to the hospital. Whenever there's a hurricane warning now, I get out with everybody else."

Even though the best defense against hurricanes now is fast detection and warning, current research in hurricane-seeding may lead to a method of weakening the storms' terrible destructive power. Dr. Noel LaSeur of the Department of Meteorology at Florida State University at Tallahassee told me that so far four hurricanes have been seeded, the last one in 1971.

The technique involves seeding the clouds just outside the eye of the hurricane with silver iodide crystals. This introduces freezing nuclei into the clouds, causing water droplets to form ice crystals. The freezing process releases heat, which disrupts the cloud wall around the eye and causes it to rebuild outward, forming a larger eye. This in turn slows down the hurricane's winds, just as a spinning skater slows down by extending the arms outward.

"If seeding experiments prove successful," said Dr. LaSeur, "hurricanes might be made much less damaging. A reduction of hurricane winds by about 20 percent would mean lowering the storm damage by 30 to 50 percent."

While weather is the state of the

Preceding pages: Motorists in Buffalo, New York, dig out an automobile, one of thousands buried during five days of snow and high winds. Beginning on January 28, 1977, the storm closed the city's schools, businesses, and factories. Snow piled up in drifts 20 feet high. Food and fuel supplies dwindled. By winter's *end, 200 inches of snow had fallen on Buffalo. On January 31, satellite photographs showed snow on the ground in all 48 contiguous states for the first time on record. The nation's bill for the disastrous winter: as much as three billion dollars in economic losses and five billion in increased fuel costs.*

*Wet-suited oysterman gathers a meager harvest in
Chesapeake Bay. The harsh winter months of 1977
brought hardship to watermen and hindered Bay
shipping. At Washington, D. C., skaters cross the rarely
frozen Potomac River. Helicopter-borne volunteers
(top) snowshoe across drifts to reach the isolated
residents of Adams, New York.*

atmosphere at a given time and place, climate is the weather that prevails in an area over a period of many years. Recently, there has been a growing worry: Is our climate changing? Are we heading into a cold era? More than anything else, the harsh winter of 1977 gave rise to this apprehension. The "Big Freeze," as it came to be called, began in September with below-normal temperatures in the East and South.

By the second week in January blizzards pummeled the Midwest. Frigid air lashed Arkansas, Georgia, and North Carolina, reaching as far south as Florida. For the first time on record, snow fell in Palm Beach and Miami, Florida. Blizzards smothered Buffalo, New York, with 126.6 inches of snow by January's end, then continued to pile up a record total of 200 inches for the winter.

"So it falls that all men are With fine weather happier far."

— KING ALFRED

In the meantime, ironically, Anchorage, Alaska, reported balmy temperatures. Hockey matches were called off because of melting ice. Bears came out of their winter sleep.

But for parts of the eastern U. S., it was one of the worst winters since the birth of the Republic. Schools closed; business hours were shortened; sweaters became a necessary fashion as thermostats were turned down.

Florida's governor proclaimed his state a disaster area because of damage to citrus crops, and Maryland's chief executive followed suit because of frozen Chesapeake Bay oyster beds and fishing areas.

A shortage of fuel was worsened by tie-ups of barge and tugboat traffic on the frozen Ohio and Mississippi rivers. Ships were marooned in ice floes on the Great Lakes. Delivery of millions of gallons of fuel oil was stalled. By February 1, an estimated two million people had been laid off their jobs because of fuel shortages. Heating bills and food prices soared.

Dr. Donald L. Gilman, Chief of the Weather Service's Long Range Prediction Group, had foreseen that the winter of 1977 would be cold, partly because of the unusually cold autumn. "We can show what happened," he told me, "but not why."

Dr. Gilman noted that by the end of September 1976, westerly winds that steer weather systems over North America had settled into a winter wind circulation earlier than usual. In milder winters, the winds blow steadily across the northern section of the continent, with gentle periodic northward and southward undulations. But in 1976 and '77, the winds swung far north over the Pacific Ocean off the California coast, carrying warm air toward Alaska. Pacific storms that normally cover the West Coast mountains with snow were breaking off the coast of southern Alaska instead. Then, as the winds crossed the Rocky Mountains, they plunged south, bringing Arctic air deep into the southeastern United States. "Such a pattern is not unusual for a few weeks during the winter," Dr. Gilman said, "but it is abnormal for it to continue for five months, as it did in 1977."

Another scientist who foresaw the cold winter was Dr. Jerome Namias, a research meteorologist at Scripps Institution of Oceanography in La Jolla, California. He predicted that the winter of 1977 would be the coldest in 20 years. "Even though the forecast was generally correct," he said, "I was a little conservative. It turned out to be much colder than I thought it would be. I don't think anybody could have predicted the intensity of the winter, although the signs that it was going to be cold were there."

What were these signs? Dr. Namias believes that changes in the sea-surface temperatures in the Pacific had a lot to do with it. As the westerly winds sweep toward North America, he explains, they are influenced by the temperature of the ocean waters.

In the summer and fall of 1976, Dr. Namias found that the central and northwestern Pacific were abnormally cool and that water off the west coast of North America was becoming abnormally warm, based on some 20,000 temperature readings taken by ships each month.

"By August of 1976," Dr. Namias said, "the water of the western Pacific was colder than I've ever known it to be − 1.6° below average."

And in the fall, he noted, warm water off the coast failed to cool. A ridge of high pressure that usually drifts south in winter remained anchored east of the warm water. The causes of these abnormal conditions, he thinks, are probably related to complex interactions between the sea and the atmosphere over the last few years. One factor was the onset in 1976 of *El Nino,* a warm ocean condition that periodically and mysteriously sweeps north along the coast of Peru and spreads westward to the central Pacific. Several scientists have speculated that this warm water may be responsible for large-scale oceanic and atmospheric changes.

The abnormal sea-surface temperatures that occurred in 1976, Dr. Namias believes, set the scene for a change in the normal wind circulation during the winter. Eventually the system began to feed on itself. "The Arctic air coming down over the Great Lakes was so cold that it produced repeated snowstorms instead of rain. The buildup of snow, in turn, reflected the sun's heat and kept the air cold. And as the warm Gulf Stream came north along the eastern coast of the U. S., it helped produce warm, moist air that interacted with the cold air over the land, creating more snowstorms."

I talked with Dr. Namias, as I had with other climatologists, about current apprehension that we may be heading toward major long-range changes in climate.

Dr. Namias believes the fluctuations to be foreseen in future weather are in the normal, changing course of nature itself. He does not foresee an imminent ice age or a global thaw; nor does he see any special portent in the abnormally cold winter of 1977.

His beliefs are based on the assumption that natural causes will continue to control the climate. But what about the impact of technology and other human activities upon weather and climate? "Heat passed into the atmosphere from the burning of fossil fuels, which produces carbon dioxide, may become an important factor in the next 50 to 100 years," Dr. Namias said. But this warming effect could be counteracted by the cooling effect of other variables, such as dust particles spewn into the atmosphere from volcanoes. This would reflect more of the sun's heat away from the earth.

Although a warming from carbon dioxide is one of the major concerns of scientists, other factors may have possible influences on climate patterns. Aerosol propellants may endanger the protective ozone layer of the atmosphere, which absorbs harmful ultraviolet radiation from the sun. Large urban heat islands could alter the temperature balance of the atmosphere.

At the National Center for Atmospheric Research in Boulder, Dr. Stephen Schneider stresses the need for prudent, long-range planning where weather is concerned. In his book, *The Genesis Strategy,* he advances a theme derived from the Old Testament. It is based on the story of Joseph, who prophesied seven years of plenty, to be followed by seven years of famine, and advised the Pharaoh to store grain to feed the people in the lean years.

Dr. Schneider says that we must hedge against possible years of bad weather by stockpiling food and fuel reserves. "I believe that the peoples of the world must build up enough climate insurance to cover themselves," he told me. "We can't predict changes in climate now. The only thing we know is that climate is variable, and the prudent man should plan ahead."

Today, a legion of dedicated scientists is working constantly to expand our knowledge of the weather machine. A great many people are engaged in weather research and experimentation. The complaint, commonly attributed to Mark Twain, that "everybody talks about the weather, but nobody does anything about it" is simply no longer true.

By Tee Loftin

Drought: Clouds of Dust, Storms of Fire

"When the well's dry,
we know the worth of water."

—Benjamin Franklin, *Poor Richard's Almanack*

During most of the 1930's, Lawrence Svobida, a young wheat farmer in Kansas, lived in suspense "looking, hoping, wishing, expecting rain that didn't come. Several times every day I stood in the hot sun to watch the sky. Time after time, clouds built up. My hopes rose. But no rain came."

Each February, cold and violent winds blew from the northwest. Tumbleweeds, driven wildly across the treeless plain, piled up against fences, trapping dust in drifts as high as the top strand of barbed wire. Stock tanks stood empty of water, half-filled with dust. Plowed soil, drought-dry, swirled and drifted against abandoned houses and granaries. Seeds that had lain in the dry soil six months without sprouting blew away. Dust filled ditches alongside roads that stretched straight to the distant horizon. Grit forced its way inside houses, into cook pots and beds. Traces of dust and wisps of winter wheat reached as far as New York and Washington.

Each July and August, hot blasts from the deserts of the Southwest swept across Kansas, burning and shriveling the bare, baked land. The sun poured on more heat. "I've known it to hit 110 degrees for several days in a row, and a few times 120," Lawrence said. "That heat killed our rattlesnakes—and it nearly killed me."

From time to time in the relentless summers, dark and evil curtains of moving dirt would blacken the horizon. Thousands of feet high, looking like horizontal tornadoes, they rolled over the land. "You got inside the house quick, watched the cloud coming, and felt it envelop you. I've known storms to last 12 hours. Dirt clicked like sleet against the window glass," Lawrence said. Thirty-mile winds, with gusts up to 60, whipped fields, buildings, and men. Occasionally, the wind would blow enough dirt into the nostrils of animals to suffocate them.

"You'd hear almost continuous thunder," Lawrence said, "and the crackle of lightning—the friction of dust particles throws off a lot of static electricity. Streaks of it ran back and forth around metal structures—gasoline pumps, for instance. Radio aerials lit up like fiery crosses. That was scary. Static electricity made the ignition systems on cars fail. That could be fatal if you got caught on the road in a storm.

"That was the Dust Bowl I knew and breathed in Meade County, Kansas." It was the Dust Bowl—measure for all American droughts—for a million families in small towns and on farms in a midcontinent belt three hundred miles wide that stretched from Canada to Mexico. It was the sort of drought suffered by millions of people in countries around the world when the rains do not come.

Lawrence hadn't made any money from his crops for years; perhaps he should leave. But in the rest of the country, the Great Depression had wiped out jobs—as John Steinbeck's fictional Dust Bowl migrants in *The Grapes of Wrath* learned in California. On the farm, at least, Lawrence had a few hundred sure dollars a year from the government. Ironically, it was for leaving a quarter of his 850 acres idle to help reduce a national surplus of wheat.

Year after year he bought bologna, canned fruit and vegetables— "gardens wouldn't grow and there were no trees"—tractor fuel, and seeds for one more try. "But some people went hungry. Families in southern Meade County had only grazing land, so there were no government checks for them.

"Finally I couldn't stay any longer. I was all by myself; my nearest neighbors, the Forst boys, were a mile away. The dust was killing me.

"Dust did kill some people I knew," said Lawrence. "One young fellow about my age spent his last day alive plowing in a dust storm, trying to roughen the topsoil to keep the wind from blowing it away. He stumbled into a doctor's office that night to be told he was dying on his feet, filled up with dust.

"Every year I saw people leaving that terrible place. Long lines of them rolled through Meade in every kind of car and truck. Some even walked, carrying what they could on their backs." In 1939, Lawrence joined the 300,000 migrants from the Kansas Dust Bowl.

"I left two years before the rains came back to Meade," Lawrence added. "Just as you don't know for a long time whether you've got a dry spell or a drought, you don't know whether the drought will disappear next month or in five years. Not even the weathermen knew. Half the time they couldn't tell us for sure what would happen today, much less next week or next month."

In his old Ford, Lawrence took a manuscript he'd written by the light of a kerosene lantern; there was no electricity on farms in the thirties. He'd had a compulsion to tell his own story of the Dust Bowl. "I wrote it all down. I said the Plains had gone to desert for good. I said I'll never go back." Another irony: In 1940, the year his gripping and now frequently cited book, *Empire of Dust,* was published, a decade of rains began, wonderfully timed for growing wheat on the Great Plains.

I found Lawrence in the crossroads community of Plunketville, in eastern Oklahoma. Now 69, his breathing hampered by emphysema, he talked about the drought and dust as personal enemies.

Reluctantly, he consented to accompany me to the Meade County farm — scene of so many bitter memories — that he had left 38 years before. In early June of 1977, I followed his tall, slope-shouldered figure through ripening, waist-high wheat in a field he had once plowed in a dust storm until he was delirious. The wind pressed billowing waves into the springy surface of the wheat.

"It's beautiful wheat," Lawrence said quietly, fingering a head heavy

FRANK JOHNSTON

Miniature wind-driven dunes entrap farm equipment in a Colorado cornfield. Owner Steve Anderson gazes across the desertlike expanse. The dust storm that ravaged his field, the nation's worst in 40 years, swept many of the plains states in February 1977.

Preceding pages: Nomadic Tuareg trek past sun-whitened bones, a common sight in the harsh Sahel region of Niger on the fringe of the Sahara. A seven-year drought killed much of the Tuareg's livestock and forced the nomads into refugee camps; in the mid-1970's government programs began providing the Tuareg with cattle, and the return of the rains to the Sahel enabled many to resume their wanderings.

JOHN AVERY

LIBRARY OF CONGRESS

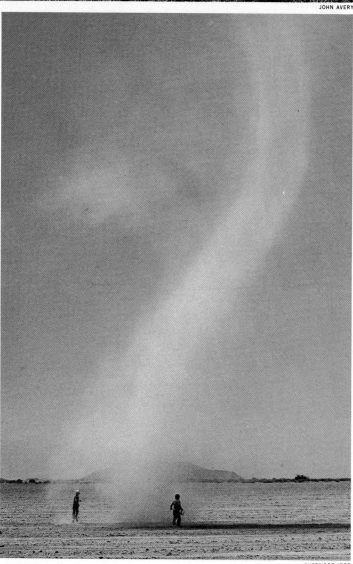

SHERWOOD IDSO

Oklahoma farmer and his sons
seek shelter from pelting grit
during the Dust Bowl. Drought,
overgrazing, and poor farming
practices brought dust storms that
battered much of the Midwest and
Southwest during the 1930's.
Lawrence Svobida (upper left)
revisits the Kansas farm he worked
for ten years before abandoning it
in 1939. Dust and drought drove
him out of the region; several
decades of more normal rainfall
have now restored the farm to lush
productivity. Boys (left) play tag
with a dust devil in Arizona. Such
tiny twisters occur when surface
heat forces air to rise rapidly.

with grain. He made a fleeting gesture with a bony hand, almost as though brushing away a memory, and turned away to gaze in silence toward the horizon. Nothing broke the distant line except small groves of trees shielding farmhouses. "Lots of trees now. Wasn't a one in the thirties," I heard him say.

"Well, I was way off when I said the Plains had gone to desert to stay," he said. "I wasn't even in the buggy. Somehow the land recovered. Kansas is green again, even after the recent 18-month drought."

How long did it take to put land stripped of topsoil back into shape, we asked Wallace McCune the next day. Ruddy and vigorous, Wallace lives with his family in a spacious home down the road from Lawrence's old place, and farms 2,000 acres.

"About two years," he told us. "Not as long as I thought it would take. In 1954, midway in five of the driest years Kansas ever had—worse than any five years of the thirties—we had a terrible dust storm. It blew out everything I'd planted. Dirt piled up in drifts like snow. We had to bulldoze it off the highway. Everything was messed up. We started working the dirt that was left to us, plowing it into the subsoil, cutting in grass, stubble, and fertilizer.

"I decided then and there I never wanted to have to do that again. Anything growing on the land will hold the soil against high winds, but growing takes water. Western Kansas doesn't have a lot of streams like eastern Kansas, and it has about half as much rainfall. In the east they started building dams and reservoirs to cushion drought when it came. Out here on the High Plains, a geological survey in 1947 showed not only natural gas but also water—lots of it—stored 100 to 600 feet down in the ground. All we had to do was pay to drill wells and pump it out."

Wallace sounded like a number of the farmers I'd talked with. "If you dry-farm—depend on rainfall—you can't stand more than three years of drought and crop failure. You'll go in debt just to eat. So you go deeper in debt to drill and buy pumps and pipes. I don't have to worry as much now about drought. I can adjust my farm's field moisture by opening and closing water valves."

"Renting rain," as some call irrigation, wouldn't have been so easy without friendly bankers and inexpensive fuel back in the 1950's. Wallace lives above one of the biggest natural gas fields in the world, but today 99 percent of it is piped east.

Neutralizing drought by tapping underground water proceeded rather slowly during the '50's and early '60's, Lawrence and I heard from County Farm Agent Eugene Harris. "But then a new generation of farmers and landowners, looking at our recurring droughts, thought it was about time for another. They speeded up well-drilling and irrigation." Gene has lived in Meade through the Fine Forties, the Disastrous Fifties, the Mixed Sixties, and the Irrigation Seventies. "By 1985," he said, "three-quarters of the land in the county where there is groundwater available will be irrigated."

Irrigation costs a lot of money—so much that Lawrence kept repeating the figures Gene gave us and adding an awed, "That much?"

More often than not, irrigated farms with modern equipment and huge new tractors complete with air-conditioned cabs carry $200,000 debt at about 10 percent interest.

"Everybody these days wants a center-pivot irrigation system," Gene said. "Push-button, they think. No dragging pipes around. It's one long pipe mounted on wheels, one end attached to the well. It can spray a thousand gallons a minute as it rolls like a clock hand over a circle of 135 acres. You don't need hired help or costly land-leveling the way you do for flood irrigation. But the well-drilling, pump, and one sprayer can cost about $80,000." In a field next to Lawrence's old place we saw a center-pivot stretched across a field.

"How much does it cost to run it for one day?" I asked. "Oh, about $85 if you use electricity, a bit less for gas. Prices have more than tripled in the past three years, and they're sure to go higher," Gene answered.

"Sounds as if you're trading the drought vampire for the fuel vampire," I ventured.

"We put it this way," Gene said. "We've solved the drought problem with an energy problem that we haven't solved."

As I left the country where Lawrence had been defeated by drought, a small dust storm beat against the air-conditioned car. It made a tiny ticking noise and I could smell the dust, musty and earthy. Ahead, I could see a dust canopy over the town of Liberal. Was the 18-month drought really over, I wondered. Or were the recent rains only a short-term break in a long-term disaster?

On the drive to Garden City, Kansas, I felt as though I were sitting in the middle of a large, flat circle that reached to the horizon. Above, storm clouds mumbled in slow baritones, and I counted five rain showers streaking the space between clouds and earth.

Just in front of me, a long cloud hung over the road. Dark with moisture, it seemed a likely candidate for a rainmaker.

"Maybe it was," Keith Lebbin told me in Garden City. Keith, who is Executive Director of Western Kansas Groundwater District Number One, took me to see the spy the rainmakers use — radar — at nearby Lakin Airport. In a small building with a six-foot dish antenna on top, I saw a large console, its twin eyes marked to measure the images of clouds.

"On this one we can tell the height of the cloud as it grows," Keith said. "And on the other screen, the intensity of the reflection shows how much moisture a cloud holds. When height and reflectivity show a cloud could produce rain if it cooled a little more, a pilot takes off in the cloud-seeding plane. He has to hurry. From birth to breakup, a rain cloud lasts only about 45 minutes. Flying below the cloud, the pilot releases sprays of silver-iodide particles. Updrafts lift the particles into the clouds. Water vapor freezes on contact with the particles. This releases heat that intensifies updrafts, making the top of the

CHARLES O'REAR

Creeping 1,400-foot-long trench menaces homes near Casa Grande, Arizona. Withdrawal of large amounts of underground water for irrigation allowed the surface to settle, opening the crack. Barely half an inch wide in 1975, it measured five feet across two years later, eroded by rainfall runoff.

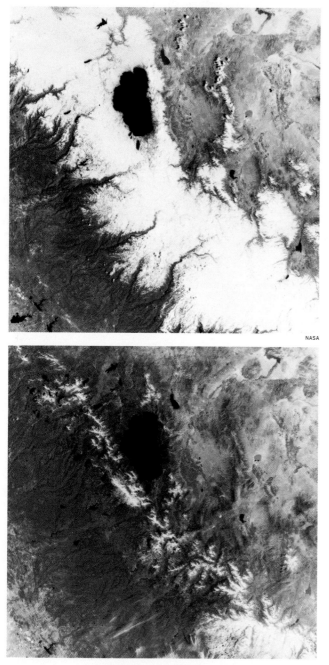

Satellite photo taken in May 1975 (top) reveals a heavy winter snowpack in a 13,262-square-mile area around Lake Tahoe in the Sierra Nevada. California farmers rely on snowmelt from the mountains to water their crops. In contrast, the same view in May 1976 shows a thin blanket of snow; it produced little melt for the drought-stricken state.

cloud go higher, where it's colder. The silver iodide particles act like ice crystals, providing nuclei on which more water vapor can freeze. They begin falling as snowflakes, then melt into rain."

How can you be sure it wouldn't have rained anyway? I asked Glenn Johnmeyer, a pilot who flew in for a few minutes' chat.

"I've been seeding clouds for seven years," he replied. "I've seen thousands of clouds from all sides. Often I've seen two identical clouds, side by side: One is seeded and rain falls, the other isn't and nothing happens. That's very convincing for me."

During droughts, he added, the trouble is that few clouds appear. "To get rain, you've got to have a cloud, and it's got to be a rain cloud," he said.

So far, no one has devised a way to herd storm clouds from one place to another—weather modification by cloud relocation—although many have tried. Indians of the Great Plains, the Southwest, and Mexico, for example, had rain gods to whom they appealed with dance, song, and magic to bring clouds. White settlers paid as much as $500 to traveling rainmakers for releasing their secret but ineffective gases during periods of intense drought in the 1890's.

In the Great Plains drought of 1909-1914, the cereal magnate C. W. Post had his men set off dynamite to try to make clouds and shake rain from them. He'd heard Civil War veterans swear that "rain follows heavy bombardments." The heavens disproved that bit of folklore as well as another: that "rain follows the plow," a sales-pitch used to lure settlers of the 1870's to the area labeled on maps as the "Great American Desert."

"There's one kind of storm—hail—we try to stop instead of start," meteorologist Edwin Boyd said when he joined us at Lakin Airport. "If you've ever seen a wheat field in June or July after several thousand tons of ice balls have pounded it, you'll know why the farmers pay to try to prevent hailstorms. Our cloud seeding technique costs about three cents an acre." I had indeed seen a field or two in Meade County where the wheat lay scattered like chopped straw.

"Curiously," Ed explained, "you seed a cloud not only to make it rain, but also to stop hail from forming."

Having lived on the High Plains for 30 years, keeping track of clouds, winds, and rains, Ed knew intimately in which direction to look for rain clouds. "Most of our water vapor comes from the Gulf of Mexico," he said, "though some of it comes from the west. But by the time those Pacific winds rise over the peaks of the Rockies, they've dropped most of their moisture on California, Utah, Nevada, and Colorado."

Why does Gulf moisture sometimes not arrive, causing a drought on the High Plains? I asked. "Well, frankly, we don't really know. There are many variables involved: temperature, atmospheric pressure, the presence of condensation nuclei. And even when they're all favorable, rainfall may cover only a small percentage of an area."

I've heard and read quite a number of theories explaining the reasons for drought. Low sunspot activity decreases solar energy and cools down earth's weather machine; volcanic eruptions throw particles into the air and screen out the sun's heat; pollutants block out—or intensify—the sun's rays; changes in earth's magnetic field alter wind and water currents; man and beast destroy grass and trees in drought-prone areas. A less-likely theory: Unfriendly nations somehow divert the jet stream with radio signals.

Some climatologists look for recurring cycles of drought and try to base predictions on them. Lawrence Svobida recalled that an expert in the thirties was certain the drought would end in 1935 because the low point of the 22-year sunspot cycle would be well past by then. "But he was wrong," Lawrence noted. "The drought got worse instead."

Climatologist Ivan Ray Tannehill noted in his book *Drought: Its Causes and Effects* that Chinese and European records of droughts during a thousand years or so show no pattern; in Africa, tree rings, Nile water levels, the layers of sediment on lake bottoms, crop failures, and other historical evidence indicate no clearly defined cycles. "Drought has been and still is one of the world's greatest mysteries," Tannehill concluded in the late forties.

Recent studies, however, seem to confirm the age-old suspicion that droughts have a tendency to repeat every 20 to 22 years, keeping step with sunspot cycles. J. Murray Mitchell of NOAA, and Charles Stockton of the University of Arizona, have used tree-ring data from around the continent to reconstruct the history of drought in the western U. S. since 1700. Their data show that widespread droughts have usually occurred in the first few years after a low point in the sunspot cycles. "I can't explain it," Dr. Mitchell says, "but this connection between droughts and sunspots is evidently not a fluke." He cautions, however, that we have sometimes had a major drought at the wrong part of the cycle, and we have also had wet years when sunspots say it ought to be dry.

Many meteorologists freely acknowledge that their 30-day predictions are still only slightly more accurate than flipping a coin. An exception is Dr. Jerome Namias, retired former head of long-range weather prediction for the U. S. Weather Service, and now an eminent researcher at Scripps Institution of Oceanography in La Jolla, California. A man of many theories, he feels confident that his latest will make predictions for an entire season of U. S. weather possible. On the basis of temperature charts of Pacific waters, he says, he predicted much of the West Coast winter drought of 1976-1977. Weathermen are watching closely to see if his prediction for a break in the drought in 1978 will be equally accurate.

There's always a region of drought somewhere in the world. It can suck the land dry for weeks before most people realize the long series of fair, sunny days is bad weather.

In deserts, the expected pattern is perpetual lack of moisture: The

Californians stroll in sunshine where fish once swam in cool lake waters. This reservoir, Marin County's largest, normally holds billions of gallons. By February 1977 months of drought had transformed it into a field of sunbaked clay. Engineers built an emergency pipeline (lower left)

MICKEY PFLEGER

LOWELL GEORGIA   DEWITT JONES

over San Pablo Bay to divert water that normally flows to southern California. Eight to ten million gallons a day flows to the stricken county. The Marin Municipal Water District encourages water conservation by distributing low-flow shower heads, dye tablets for detecting leaks, and plumbing-repair information. Officials also charge penalty fees to customers who exceed allotments. His water crisis solved, Doug Keister (left) of Oakland wets down his trees and shrubs. Keister drilled down 20 feet and brought in a well in his backyard. It produces 1,200 gallons of water a day.

little rain that falls comes in sporadic, violent storms. Nearly all deserts lie along the 30° line of latitude, both above and below the Equator. Winds blowing in a great oval pattern above the ocean and nearby land drop their moisture before they reach the final turn of the circle.

Lands on the fringes of deserts contend with periodic drought in a mixed climate: many normal years for growing crops, a few years of exceptionally heavy rains, and occasional rain failures.

Drought can extend over a country or an entire continent. Australia has experienced continent-wide drought about twice a century. In the past 50 years, most places in the world have experienced drought. In a few places, such as the African Sahel, several years of continuous drought, intensified by overgrazing and population growth, have wrecked the economy and lives of the people.

Sometimes when an area endures a drought for five or ten years, fears rise that desert conditions will take over permanently. A new United Nations Committee on Desertification now watches long droughts and brings funds and technology to fight the threat. So far, rains eventually have resumed their regular patterns in the drought areas.

Since the Stone Age, whenever the weather has turned dry and hot and windy for days on end, people have diverted streams to water their crops. Today, the wealthy and technologically advanced nations invest heavily in engineering projects capable of neutralizing drought years.

James Power, Director of the Kansas Water Resources Board in Topeka, showed me on a map in his office the 20 big and 500 small dams and reservoirs in eastern Kansas, all built since 1957. They cost more than a billion dollars to construct. "They can hold the line against drought in eastern Kansas for three years," Jim said. "After that, a drought would mean an emergency situation for this area—unless we build more water traps and reservoirs."

On the drier side of the state, where Lawrence Svobida suffered

through the thirties, a three-year rain failure in land irrigated with underground water would hardly have an effect in the next decade or so. "But an economic Doomsday could come in 10 to 25 years," Ed Gutentag told me. "By 2020 or so it will cost so much to produce gas that it will be uneconomic to pump water for irrigation in areas of heavy use."

Ed, a happy transplant from the Bronx, heads the U. S. Geological Survey office in Garden City. He nodded vigorously when I told him Gene Harris's remark about solving drought problems with an unsolved energy problem.

"One group of farmers in Wichita County has started to work on that," he answered. "They're cutting the amount of water they use in order to cut gas use and save money. That might stretch the underground supply of water and gas, a partial solution to give the technologists time to find new fuel sources. For the first time since they began irrigating, these farmers are metering how much water they use. Ordinarily farmers say they 'just know' when to water, and leave the pump running until the land 'looks wet enough.' So corn that needs one foot of water per acre twice a season often gets three or four times that much."

Recharging aquifers, the underground reservoirs, is the next step. Wallace Robinson, who farms near Scott City, Kansas, has already started. "His farm, close to 4,000 acres, is dotted with ten-acre lakes after a heavy rain," Ed said.

"Three weeks later, the water's gone," Wallace explained when I talked with him later. "Tests show that water will easily seep into the aquifer—provided the recharge area doesn't get clogged with silt and bacterial slime. We disk it after each soaking, and that opens the soil to the seeping water. The ponds are only six inches lower than the adjacent 30 acres that they drain; dikes hold the water in bounds. I know the aquifer is recharging because the water table has dropped less beneath my farm than beneath neighboring farms. In a few years, I think

In the glow of a tungsten lamp, a miner clears loose rock from one of the tunnels that will divert water from Colorado's western reservoirs to the state's dry eastern slope. The network will carry billions of gallons yearly.

I might have a stable water table."

I tried to calculate how fast he could fill the aquifer if his hundred ponds could draw water from rivers now gushing out to sea, unused. Numerous plans exist for diverting such water to the Great Plains and the fertile but dry states west of the Rockies. The most spectacular calls for another Alaska pipeline, to carry water from the Juneau Icefield and the Yukon River.

When I left Kansas, clouds from the Gulf were bringing rain every week. Mile after mile of golden wheat awaited the harvest; corn grew knee-high and crisp. Surely the long drought was ending.

But west of the Rockies, the drought was intensifying in the summer heat. For the second winter in a row, few Pacific clouds had blown in during the wet season. Summers there are normally dry and hot, but after a whole year without rain the drought was pressing deeper into the land and its browned-out grass, into its reservoirs, and into its brush and timberlands.

Lightning began to scratch the sky in late July. By mid-August, it had struck earth nearly 4,000 times

to start as many forest and range fires in the West. Most were quickly put out, but a few burned more than 400,000 acres in northern California, Oregon, the northern wilderness of Idaho, and the mountains of Utah.

Alaska, also feeling drought, lost about two and a half million acres of timber and tundra to fire in 1977. In the last days of July, as I flew north from Fairbanks to Barrow, I looked down at a vast and billowing cloud of thick brown-gray smoke from the still-burning wilderness.

A few days later I checked in with Arnold Hartigan at the Boise Interagency Fire Center in Idaho. Outside his door a double line of young fire fighters from all parts of the United States streamed past us into classrooms for special training. "There are 600 in the barracks waiting to fly out to fight fire," Arnold said. "We've had close to 10,000 through here in the past three weeks—including quite a few women. Some helped save houses in the hills around Santa Barbara, California. Some fought the big fire in Los Padres National Forest that damaged the Carmel watershed."

The Fire Center monitors the

JOHN AVERY (ABOVE). KERBY SMITH

Circles of lush corn grow where
sprinklers wheel in fields near
Garden City, Kansas. Center-
pivot irrigation, widely used in
the United States since the early
1950's, draws heavily upon
underground water. Each
sprinkler, a quarter of a mile
long, covers some 135 acres. The
cost of one well, pump, and
irrigator: approximately $80,000.
In a field near Wasco, California, a
combine cuts and threshes barley
stunted by drought. The severe
water shortage in the San Joaquin
Valley prevented the farmer from
irrigating. The stalks yielded less
than half the grain of a normal year.

country's forest and range fires and dispatches trained fighters, helicopters, bulldozers, planes, and equipment from around the country to work on big blazes whenever local fighters call for help. "Lightning starts about 50 percent of the fires in the West," Arnold said as we watched teletypes and facsimile machines pour out news of weather and fires.

"Until this year, we had no way of knowing where lightning had struck in a remote area—until the blaze sent up enough smoke to see. Often by that time it was a big fire, difficult and costly to kill," Arnold said. "Now we have an apparatus that picks up the magnetic field impulses created by a lightning strike. Electronic equipment displays them on a plotter and tells us where to send a plane to look for fire."

We visited the Center's control room, where dispatchers at three large telephone consoles rounded up fire fighters, planes, and equipment to send to danger spots. Later, in the communications shop, I talked with Phil Sielaff. "He's one of five men in the world who has installed lightning detectors," Arnold said. The towering young blond had just completed work on the first three in the West—at Susanville, California, at Vale, Oregon, and at Elko, Nevada. "So far we can only locate the square mile within which lightning has struck," he said, "but soon, by triangulating data with a computer, we should be able to pinpoint the strike area much more precisely. Eventually, 16 stations will cover most of the West."

The lightning antenna, as Phil calls it, looks like a nine-foot tepee made from four pieces of ordinary plastic sewer pipe. Two loop antennas are housed in each leg, and can pick up signals from 300 miles away. Phil told me that Alaska, which has millions of square miles of forest, tundra, rangeland, and brush, asked for such a device, and Dr. E. Philip Krider of the University of Arizona was asked to develop it. "So the first units went to Alaska," Phil said. "The devices have been a great help to the state. Officials have been able to cut out about half of their detection flights and by early action have saved a lot of timber besides."

How about the millions of acres I saw burning? "Alaska is suffering from the same lack of moisture as the rest of the West," Phil said. "The foliage, especially the floor litter, gets so dry that lightning strikes can cause many rapidly spreading fires. Strong winds will keep them rolling. Smoke boils so thick that the pilots can't see well enough to drop chemical retardants accurately."

What will happen in the summer of 1978 if the drought continues? "The potential for another severe fire season would be high," Arnold said.

In California, Oregon, Nevada, and Utah, worried officials were calling the drought of 1977 the worst in Western history, though the one in 1917-18 was perhaps comparable. There had been no winter rain in the valleys, and little snow on the mountains. I visited Yosemite National Park in California's Sierra Nevada—Spanish for "snowy range of mountains"—to talk with rangers who periodically climb the slopes to measure snow depth and wetness. The April weather was ominously like summer in that beautiful high meadow.

Ranger James Sleznick, Jr., gave me the end-of-season reading: "Usually at 7,000 feet we have an average of 78 inches of snow, which will melt to about 40 inches of water. This year there was only 14 inches of snow, or six and a half inches of water. And it's that same proportion all the way up the heights."

At the foot of the Sierra in the San Joaquin Valley, I drove past miles of Fresno County vineyards and fruit trees, and green fields of vegetables, all with irrigation water standing in furrows. Beside the lush fields lay desolate acres brown with thirst. I crossed the Lakelands Canal, which brings water from Pine Flat Reservoir 30 miles away. Farmers would be permitted to take only a quarter of their usual water allotment from it this year. Outside a mammoth co-op gin, bales of cotton and a wall of hay sat uncovered, as if no one ever expected it to rain again.

At Wasco in Kern County, I stopped at a cafe where farmers and ranchers talked water and crops over their coffee. Tall, robust E. V. Gordon, whose factory in Bakersfield makes alfalfa pellets for cows and chickens, remarked that Kern County has changed in less than ten years from a desert of tumbleweeds to a farmland richer than that of the Imperial Valley. "Big companies drained off the oil, then sold the land to farmers who drilled for water. There's a lot underground, so we aren't totally dependent on canals and reservoirs. Droughts won't kill us as long as we can get electricity."

But the pumped water costs four times as much as canal water. And pumping will become more expensive as the cost of hydroelectric power increases, due to lower water levels behind dams. The price will be passed on to American consumers, who get 40 percent of their fruit and vegetables from California.

In San Francisco, the first major city in the U. S. to impose voluntary water restrictions on its residents, my friend Helen von Ammon didn't find the cut inconvenient. "Do you remember that old song, '*Oh, it ain't gonna rain no more, no more, it ain't gonna rain no more! How the heck can I wash my neck if it ain't gonna rain no more?*' Well, I've been taking that seriously for months. I cut baths to fragmentary showers. All junky water I save for houseplants—they never had it so good, and there's a surplus waiting for them. I flush the toilet twice a day and machine-wash dishes once a week."

Edward S. Ross, Curator of Entomology at the California Academy of Sciences, wondered how an individual could possibly use as much as 45 gallons of water a day, the average allowance in Marin County, much less neighboring San Francisco's limit of 52 gallons.

"Of course I know," he smiled wryly. "We run most of our water down the sewer to the sea and won't take conservation seriously until dust comes out the other end. Did you know we use more than half the
(Continued on page 154)

CHARLES O'REAR

Physicist Ray Jackson of the U. S. Department of Agriculture takes the temperature of cotton plants in Arizona through an infrared thermometer. The instrument helps measure how much water farm crops need, thus avoiding waste. Water witch R. C. W. Frisbey, at a drilling site near Wendover, Utah, shuns technology and seeks water with a metal divining rod.

DAVE CONLEY

JIM BRANDENBURG (ABOVE). PAINTING BY LISA BIGANZOLI

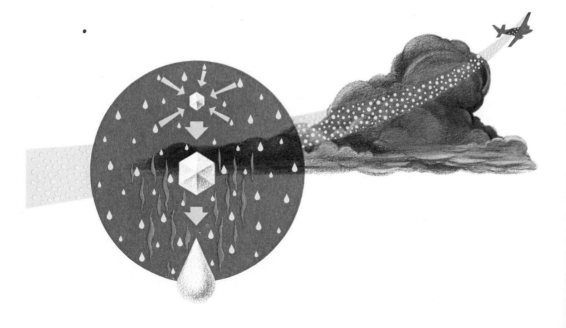

Cloud Seeding: Dispersing Hail and Inducing Rain

Wingtip generator and a battery of flares scatter silver-iodide crystals beneath clouds heavy with hail; updrafts carry the crystals into the clouds. Meteorologists in North Dakota maintain that they successfully diminished damage from the July 1977 hailstorm that bombarded wheat fields. Scientists seed clouds to induce rain as well as to disperse hailstorms. In the painting at left, a plane sprays silver-iodide particles (blue dots) into a cloud of supercooled water droplets. Each of the trillions of particles (insert) provides a nucleus for surrounding moisture (tear-shaped drops) to freeze on, forming an ice crystal. The crystal falls earthward as a snowflake until it reaches warm air, where it becomes a raindrop.

149

hampering fire fighters. Rubble surrounds a youngster's charred motorcycle. Jere Steele (below, right) worked with a fire-fighting team trying to control the blaze by helping to dig a firebreak around it.

AL MOLDVAY (ABOVE). JAMES A. SUGAR

153

Summer rain clouds, captured in a panoramic view, glide toward Phoenix, Arizona. Such clouds often warn of violent dust storms called haboobs. Three hours after the photographer made the series of pictures at top, a thousand-foot-high wave of sand and dust buffeted the city (lower pictures). Such onslaughts sharply reduce visibility, making driving hazardous. Haboobs, lasting as long as six hours, owe their existence to moisture. Arizona's summer thundershowers cool the air, creating powerful downward air currents that carom off the desert floor, churning up sand in front of the approaching downpour.

water in our homes flushing toilets? And this doesn't even include the water we use watering gardens and washing cars. Maybe this drought will be a blessing in disguise. The Marin Municipal Water District has given away more than a hundred thousand displacement bottle kits for residents to put in their toilet tanks to cut down the amount of water used, and California has a new regulation that after 1977 builders must install toilets with tanks designed to use less water. And all because most rainfall fronts have been deflected northeast around a stationary high pressure area off the California coast."

I drove across the Golden Gate Bridge to Point Reyes, a narrow, 280-foot-high peninsula jutting into the Pacific, to visit the drought-stricken dairy farm of George Nunes. I saw a pond with a bottom of cracked mud and only a puddle in the middle. Water for the cows was hauled from

seven miles away. "It takes 10,000 gallons a day for 275 cows to make six gallons of milk each," George said. "We wash the udders and teats, the milk tanks and machines, but the Health Department knows we don't have the water to bathe the cows and keep their holding areas washed.

"For me," George went on, "the drought last spring meant two extra months of buying hay for my milk cows — at $16,000 a month — because the pasture grass didn't grow. This year may be better. I've put in a crop of oats which will grow with whatever rain we get, plus the moisture from ocean fog. That will give me an ample amount of silage — I hope."

In the industrialized countries of the West, each person uses an average of 70 gallons of water a day. During a drought, the shifts in daily routines for people who seldom think twice about turning on a faucet sound much the same, whether they're in San Francisco or London. But the flavor is different. Back in the summer of 1973, for example, England didn't get much of its usual drizzle. "It happens about once every 40 or 50 years," said a Water Resources spokesman, "and it will go away."

But it didn't go away, and by the summer of 1976 had turned into a severe drought. Marion and Arthur Boyars, London publishers and long-time friends, sent me their inimitable summary of that summer, drawn from newspaper headlines.

"In June they were saying 'Worst drought in 100 years,'" they wrote. "Standpipes — faucets to you — in the streets, tankers hauling water to rural areas, heath fires in Somerset. Greatest shortage of potatoes in memory. They cost more than Mediterranean oranges.

"July: Passersby cheer lady in Chiswick for walking the street cool and naked. Pleasure boats ordered off the canals. The Thames looks sick, low, muddy. But no stink: New sewer completed last year.

"August: Worst drought in 250 years. Water cut off overnight in South Wales, center of drought. Authorities with power to fine water-wasters accused of sending spy planes to spot green gardens. Tempers rise: Welsh angry at continued export of their water under old contracts. Call for a national water grid. Queen decides to let royal gardens die. London's fountains turned off. Water pressure reduced, taps run slow. Clouds of smoke from burning forests and heaths. Drought grips all western Europe. Shortage of beef, wine, milk saves Common Market huge price subsidies. Drought worst in 500 years. Rain one-quarter normal. Rainy London day makes the news on the telly. St. James's and Hyde parks are like deserts. East Anglia a mini-Dust Bowl. Water-diviners do a roaring business.

"September: Wales, usually wettest outpost in Europe, worries about cutback in factories of big water users—American industrial chemicals, plastics. Suggestion for stretching water: dilute it. Life goes on as usual, actually more fun than usual because of fine weather. Water for homes cut off in parts of southwest England.

"Mid-September: Finally! Torrential rains! Six months later: Rain still falling. English again living under umbrellas."

Such jaunty details are not available from the Soviet Union, whose unirrigated prime farmland is among the most drought-plagued in the world. "But Soviet weather stations regularly exchange weather data, under the direction of the World Meteorological Organization, with all the countries that want it," I heard from meteorologist Malcolm Reid.

In Washington, D. C., Malcolm heads a division of the Commerce Department's Center for Climatic and Environmental Assessment. The center provides worldwide drought information, as well as data on the impact of climate, to the federal agencies that are concerned with food production, energy consumption, and

disaster relief. "We also have weather satellite pictures of cloud cover. Of course they don't tell us for certain whether the clouds are producing rain. To know that, we need data from a weather station."

A wall-size map of the world, with major wheat-producing areas marked in red, is the center's scoreboard for adverse crop conditions. "Drought is by far the greatest weather enemy of crops," Malcolm said. Magnetic blocks representing droughts freckled the continents. Most of the blocks sat in major wheat-growing areas.

"Wheat doesn't require a great deal of water, and it thrives on plains that stay fertile because nutrients in the soil have not been leached out by heavy, frequent rains," Malcolm said. "Drought in those wheat areas hits hard at the local—and some-

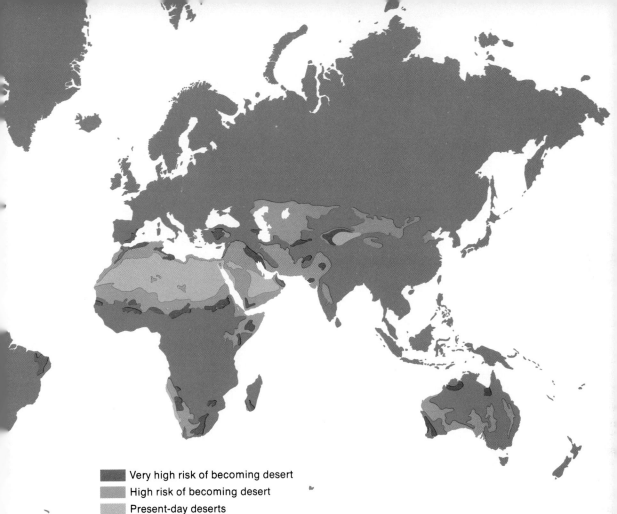

Very high risk of becoming desert
High risk of becoming desert
Present-day deserts

MAP BY PETER J. BALCH

times the world—food supply. In the future, the effect of drought is bound to be multiplied by the increase in the world population. India's yearly growth, for example, is equal to Australia's total population."

Malcolm brought out a series of maps to tell me the story of the devastating drought in the Soviet Union in 1975, even more disastrous than the one in 1972. It began in April as a small spot near Kuybyshev, between the Volga River and the Ural Mountains. By the end of June, numerous drought spots had merged to become a large area the shape of some prehistoric monster. It extended across the fertile triangle formed by the Baltic, Black, and Caspian seas where 90 percent of the Soviet Union's farming, industry, and people are located, and covered the great plains east of the Urals. It was January before

Restless Deserts Grow Ever Larger

Red alerts and orange warnings in this modified United Nations map show the scope of an environmental danger: creeping deserts — or desertification — that threaten large fertile regions around the globe. Deserts spread for several reasons. Overpopulation forces millions of farmers onto marginal lands; growing herds of livestock overgraze the vegetation; people strip the land of trees for firewood. Experts believe a decrease in population pressures, conservation of soil and water, and better livestock management will reverse the trend.

Servants of a Tuareg chief draw water for his herds at a well in Niger (above, right). Later in 1973, the chief lost thousands of cattle to the Sahel drought. It swallowed up wells, denuded pastures, and drove a Tuareg mother and her children (right) into a sub-Saharan refugee camp near Niamey in Niger. Relief workers at the Niamey camp cared for thousands of nomads during the drought; many, weakened by malnutrition, died of disease.

In Brazil's sertão, a semiarid region in the northeastern bulge, weather patterns change every few years, shutting off the normal seasonal rains. A family of Sertanejos (above, left) interrupts its flight from the dry backlands to fill gourds and to fish shrinking ponds at the village of Inhamuns.

A potential threat all over the world, drought strikes especially hard at peoples living near deserts, where rainfall—always scarce—sometimes fails for months on end.

snow fell, finally ending the drought.

The Soviet Union bought millions of tons of wheat that year from the United States, Canada, and Argentina. American farmers reaped a bonanza, and the price of bread worldwide bounded up, nearly doubling in places.

"Almost every other year between 1961 and 1975 drought has cut back the Soviet wheat crop," Malcolm said. "The rain clouds that water the Soviet Union come from the Atlantic, nearly 2,000 miles from Moscow, 3,000 miles from the wheat fields behind the Urals. We're lucky to have the Gulf of Mexico relatively close to the center of our continent."

Since 1949, the People's Republic of China has transformed the northern half of the country from one of the worst drought areas in the world into one of the more successful drought-resistant regions, according to Charles Liu of the U. S. Department of Agriculture, a specialist in the economy of Chinese agriculture.

Even as late as 1959-1961, severe drought caused widespread crop failures. So the Chinese increased the emphasis on fighting drought with flood-control and irrigation facilities as the only answer to producing the larger crop yields needed to feed the growing population.

During the winters, millions of men and women have been doing construction work on reservoirs, canals, deep wells, and dams to control the floods of the Yellow River and to provide irrigation water for farms. When drought struck in the spring of 1977, the Chinese Government called for an all-out effort to fight it, including the formation of long human chains in bucket brigades.

Today, about 40 percent—roughly 100 million acres—of China's cultivated land is irrigated, and the reported goal for 1980 is 50 percent.

In neighboring India, where millions once died in times of drought, nearly all wheatlands are now irrigated. Gifts of food and credit tide the country over when the monsoons fail for more than one season, as they did in the early 1960's.

From Niger, in the Sahel region of Africa, Dr. John A. Dreisbach wrote me his eyewitness account of the six-year drought that reached its peak in 1973. That year, Dr. Dreisbach, a medical missionary who has lived in Africa for 30 years, moved his family and clinic to Niger, one of the world's most drought-prone countries. "I saw thousands of cattle carcasses along the way. Nomads who once owned them were in relief camps eating gift food sent by the rest of the world. We saw much malnutrition and sickness. In their weakened condition, many people died of pneumonia, measles, and whooping cough. I saw little starvation as such." The

---

"Here I am, an old man in
  a dry month,
Being read to by a boy,
  waiting for rain."

—T. S. ELIOT, *Gerontion*

U. S. Department of State estimates that 100,000 died in the region during the drought years.

Lying beyond the southern edge of the Sahara is the Sahel, a region of scant rainfall that runs from Africa's west coast inland for 2,000 miles. A great deal of it resembles the lightly grassed plains of Kansas, even to having dust storms called haboobs, Arabic for "violent winds."

Normal rains began returning to the Sahel in 1974. "The land has come back remarkably well, more rapidly than I expected," Dr. Dreisbach wrote. "Good crops and pastures where nothing has grown for several years. Cattle, sheep, goats, and camels are increasing in number. Many of the nomads are back in their traditional grazing areas, some leaving regular work and pay. Great activity in the big market at Abala. I was surprised at the ready cash." Old-timers remember other severe droughts, Dr. Dreisbach said, and look back on this one as just another. It was not so bad as the drought of 1910-1913, when no help came, and many hundreds of thousands died.

Dr. Dreisbach mentioned that

new wells had been dug and old ones cleaned and deepened so that long-abandoned gardens were irrigated again. How much water lies below the Sahel? A great deal, I learned. Hydrologists believe that large quantities of water are stored in the sandstone beneath the land, some of it within 100 feet of the surface.

"You see, the water is there," I heard from Robert P. Ambroggi, a Corsican who has spent a large part of his life working on the world's water difficulties. He now advises the United Nations and the World Bank on hydrology. "I contend that mankind should deliberately draw down the underground reservoirs far more than we already do to meet the needs of agriculture, industry, and community—especially during droughts.

"Experience shows," he said fervently, "that the heavy rainfall of every decade's 'wet year' in drought-prone areas will replenish an aquifer sufficiently to last 10 to 15 years. I've seen it happen in Tunisia and Morocco. We must see aquifers as both a source of water and as a place to store it. To have storage space, you must empty the aquifer."

Dr. Ambroggi mentioned several other places—India, Israel, California—where aquifers are now emptied and refilled regularly. "In the Indus River Valley of Pakistan, site of the largest irrigation system in the world, canal and field seepage filled the aquifers during the century after 1860. That water is now pumped out to help expand the irrigation system. This is one area in the world where drought has no effect."

Dr. Ambroggi leaned across our lunch table at the World Bank in Washington to speak urgently about water problems on a worldwide basis. "In 1972," he said, "the water deficit of 400 cubic kilometers that caused a food crisis was minuscule compared with the 45,000 cubic kilometers of water available in the top 100 feet of groundwater reservoirs. From those reservoirs, a cubic kilometer of water can be obtained for $20,000,000. The same amount stored in a river-dammed lake costs $100,000,000. The economics alone should interest governments of the world.

"Water is basic, and in the future, with population increases, we will have droughts created by human overuse of normal water resources. Water must become a matter of national policy and action—the sooner the better."

Experts, however, disagree about the amount of water that can be pumped out of an aquifer without adversely affecting the groundwater level of a region. Some doubt that rainfall would be able to replenish an aquifer drawn down too far. They note that in the 65 years since the drought of 1910-1913, the rains still have not replenished Lake Chad in Africa to its pre-drought level.

Dr. Ambroggi acknowledged that better methods are needed for recharging the aquifers. "Rain catchment pits to collect runoff hold great promise. They're cheap and easy to maintain, great for the poorer countries like those of the Sahel. And if clean water is piped from distant rivers to drought-prone lands, it could be easily pumped into a deep well without clogging the porous stone."

Perhaps one day the cost of removing salt from ocean water will be low enough, and the volume large enough, to justify piping water inland to drought-prone lands.

As I left the World Bank, I thought of Wallace Robinson, the recharge pioneer of western Kansas, with his rows of ten-acre ponds, and the visionary plans of engineers to pipe water from Alaska. But where were the funds to come from for all that—and for the well-drilling, pumps, fuel, and irrigation equipment on each farm?

I had a haunting recollection of the concerned expression of Kansas farm agent Gene Harris when he linked the energy crisis to drought-proofing. Nevertheless, the excitement that Dr. Ambroggi's ideas had stirred would not quiet. "The water is there"—underfoot if not in the heavens, and one day the nations of the world could, if they put mind and money to it, alleviate the problems of drought—and perhaps even solve them.

By Cynthia Russ Ramsay

Water: Time's Relentless Sculptor

"Every valley shall be exalted, and every mountain and hill shall be made low: and the crooked shall be made straight, and the rough places plain."

—*Isaiah 40:4*

There was no escape for the people in the Covered Wagon Cafe when rampaging waters tore the building from its foundations and swept it into the night.

Nothing was left of the Waltonia Motel; only bodies remained, of people who were on the roof when the building splintered and disappeared.

Rescue was impossible for the families trapped inside cars and trailers and carried away by the current. Days later, in the wasteland of mud and silt left by the flash flood, the vehicles were recovered—twisted, crumpled hulks buried as much as six feet beneath the bed of the Big Thompson River.

In the deluge that thundered down the canyon, 139 perished—some drowned, but most were battered by debris and boulders. At least six missing people remain entombed somewhere beneath tons of sediment flushed from the slopes and the channel of the river.

The disaster struck on the evening of July 31, 1976, when Colorado's Big Thompson and its North Fork, swollen by torrential rains, turned the cottages and campgrounds along a 20-mile length of canyon into graves. "People are still living this nightmare," said Mrs. Dorothy Venrick, the postmaster of Drake, a mountain hamlet that once had a population of 200; Drake lost 40 of its citizens to the fury of the river.

"It had been raining very hard for about an hour when I heard this tremendous boom, like the roar of a low-flying jet. I looked out and there was a trailer tumbling down the river, bobbing in the waves like a toy in a bathtub. Cars too! I don't know how many I saw wash away. Cars, homes, trailers—they just kept on coming.

"You can't imagine how helpless we felt, but the river was so wild—churning and foaming. There was nothing we could do. Nothing. It was dark, mind you; pitch black until the lightning struck, and then you wished you couldn't see." Mrs. Venrick shuddered and looked away.

Nine months after the flood, her eyes still spoke of her ordeal and her grief. We were sitting around the big table in her kitchen. The road, which had been rebuilt, and the river, a pleasant mountain stream, were 100 feet below the house, one of half a dozen buildings in Drake that lay above the reach of the flood.

James Venrick, her husband, broke the silence. "We could hear the propane tanks exploding over the roar of the water. The air reeked of the gas. Every time the lightning flashed, we could see the propane, a blue haze swirling above the waves."

The storm drenched the surrounding mountains of the Front Range with more than ten inches of rain in four hours, and the water surged down the steep slopes. It tore topsoil away and turned gullies and ravines into wild rivers that dumped water, sediment, and debris into the Big Thompson and the North Fork. It was a classic example of a flash flood in the mountains.

Mrs. Venrick, echoing the reactions of so many survivors, said softly, "It didn't seem real."

The violence of the water was indeed hard to believe. It smashed a two-story municipal power plant at the foot of the canyon into a pile of brick rubble, and reduced a concrete bridge to gravel. It chewed U. S. Highway 34 into pieces of asphalt, bulldozed houses, and in a few hours turned Tom Hart's tomato patch into a ten-foot-deep ravine strewn with three-ton boulders.

Even more astonishing were Dr. Donald Doehring's figures on the tremendous amount of energy in runoff. A geologist at Colorado State University in Fort Collins, and my guide along the Big Thompson, Don had calculated that an inch of rainfall draining one square mile of land and descending a scant 1,000 feet has the energy of some 60,000 tons of TNT.

When Don and I left the Venricks to trace the river's trail of destruction upriver to Estes Park, Mrs. Venrick's parting words were spoken in earnest. "I pray no one has to go through anything like that again."

Don waited until we were out the door to comment. "Floods will occur no matter what man does. They are natural events in the life of a river. Unfortunately, that's sometimes the price people must pay for living on a floodplain." As we drove along the road that twisted through the narrow valley, uprooted trees and wedges of raw earth, where the riverbank had caved in, marked our route. The river had left scars everywhere.

Don pointed to the forested slopes and granite crests that walled the valley and said, "To understand what water can do, consider that it leveled the Ancestral Rockies — once a mountain mass perhaps as mighty as the Rockies of today. And over millions of years, unknown rivers and streams carried the spoils of sand, gravel, and clay east to form part of the High Plains. Think about that for a while," he said.

And I would, for it was part of the awesome saga of water — a force that shapes the surface of the earth — which had brought me to the Big Thompson in the first place.

But this disaster, like the flash floods on the Big Thompson in 1921, 1945, and 1965, was just a turbulent moment in water's assault on the land. For water, in the splash of a raindrop, in the flow of a river, in a crystal of ice or a great glacier, in the crash of waves against the shore, is grinding and dissolving rock, ceaselessly carving valleys and coastlines.

At a shallow stretch of river, we stopped beside a small grove of aspens still bare at the beginning of May. The water eddied around the boulders and lapped at stones worn smooth by hundreds of thousands of years of friction. On the bottom we could see pebbles rounded by rolling along the streambed and polished by countless waterborne grains of sand. In a deeper channel upstream, the water glided into a quiet pool, black

Flood victim Ernest B. Ray leafs through a mud-soaked family album in his home near the Levisa Fork of the Big Sandy River in Pikeville, Kentucky. Water reached the roof of the house after heavy rains in April of 1977 forced the river out of its banks.

Preceding pages: Muddy waters of Hurricane Agnes flood Harpers Ferry, West Virginia, in 1972. Each year in the U.S., floods take a greater toll of life and property than any other natural disaster.

Shopowners sift through merchandise ruined by waters of the Tug Fork of the Big Sandy River in Williamson, West Virginia. Rising more than 30 feet in 12 hours, the river cascaded over a floodwall, destroying or damaging most downtown businesses. In nearby Matewan, milk cartons filled with drinking water line Main Street (below). Residents of Williamsburg, Kentucky (opposite, lower), reach the business district in boats. Here, the Cumberland River flooded during the same three-day downpour in 1977 that sent several rivers in the area over their banks. Floodwaters left a wake of property damage totaling hundreds of millions of dollars. The Levisa Fork (opposite, upper) winds for 160 miles through the Cumberland Mountains. Here, high water reaches riverbank trees at Pikeville, Kentucky.

MICHAEL COERS (BELOW). NATIONAL GEOGRAPHIC PHOTOGRAPHER JOSEPH H. BAILEY

and glossy in the amber light. The Big Thompson, less than a dozen feet wide, seemed grossly misnamed that serene spring afternoon.

"The transformation of rivers like this into killer floods takes place hundreds of times a year," said Don.

Dr. Eugene L. Peck, Director of the Hydrologic Research Laboratory at the National Weather Service in Silver Spring, Maryland, talked with me about the conditions that cause rivers to rise. "Heavy rains, of course, are responsible for many floods, but in some parts of the country, such as the Northeast, the Mississippi and Ohio valleys, and around the Great Lakes, the spring thaw takes first prize for causing trouble.

"There are many factors that affect the flood potential of the spring thaw," he said. "For example, some soils can hold more moisture than others. Then the wind, humidity, and temperature, which determine the rate of evaporation, also control the amount of water in the runoff. But the crucial thing," he added, "is the water content of the snow."

I knew that Rocky Mountain powder holds less moisture than the heavy, wet snow of the East. On the average, a foot of snow in the Rockies contains a little more than an inch of water. But the River Forecast Centers of the National Weather Service, which are responsible for issuing snowmelt flood forecasts, need specific information about the volume of water produced when snow melts.

"In the heavy-snow areas of the western mountains," Dr. Peck said, "the amount of water is determined by surveyors. In the flat areas of the northern Plains States, we are beginning to use aerial measurements of the natural gamma radiation to get better estimates. Since the earth releases gamma radiation at a steady rate, we can measure the amount emanating from a specific area before it snows; then after a snowfall we fly over the same area and take more readings. From the difference we can calculate the amount of water, since a blanket of snow decreases radiation."

But heavy rainfall still causes most floods. A downpour hit parts of Virginia, Tennessee, Kentucky, and West Virginia in April 1977 when the land was still waterlogged. After five inches of rain, the Tug Fork of the Big Sandy River began to rise, and in Matewan, West Virginia, reached as high as the rafters of Pastor Wayne Lairson's church.

Members of Company D, 1092nd Engineers of the West Virginia National Guard, who were helping the pastor and the rest of the town clean up, had just found some fish in the church basement. "Just little fellows—nine suckers and two sand pike," said Staff Sgt. Barry Cantrell.

"It's hard to describe how the town looked when we first arrived," he said in reply to my question. "But if you've ever been on a farm and seen a hogpen full of mud, it might give you some idea."

We were shouting to be heard above the rumble of bulldozers leveling debris and ruined buildings. Twenty-eight houses had been washed off their foundations.

Down the street, Benny Accica was working in the wreckage of his grocery store, trying to salvage what he could. Cases of soft drinks—the bottles encrusted with silt—were piled outside. His daughter Pauline sighed. "I wonder if we'll ever get rid of all the mud. Daddy lost a shelf or two of merchandise in '53, '57, '63, and '69, but those floods were nothing like this. The water came up so fast nobody had time to do anything."

It came as no surprise to me to learn that flooding is the most widespread geological hazard in the United States, and that fully half the communities in the nation face the danger of water breaking out of streams and rivers. The American Red Cross provided assistance to victims of 1,456 floods in the years between 1952 and 1977.

Around the world, floods inflict a greater loss of life and property than any other natural disaster. In the U.S., floods cause an average of more than a billion dollars' damage a year.

The statistics of death are more tragic. History's most lethal flood occurred in the autumn of 1887, when China's Yellow River inundated an area as large as Alabama, killing some 900,000 people.

In spite of the funds and technology employed by engineers to blunt their fury, rivers continue to go on the rampage, taking a grim toll of lives. In Johnstown, Pennsylvania, for example, where the flood of 1889 claimed more than 2,200 victims, and the flood of 1936 took 25 more, the U. S. Army Corps of Engineers spent millions channeling streams and installing concrete revetments along riverbanks to make the city safe. But in July 1977, tributaries of the Conemaugh River once again gushed down from the Allegheny Mountains, engulfing the city and its suburbs and taking some 85 lives. Two months later, Brush Creek, a stream that trickles through Kansas City, Missouri, spilled out of its banks, killing 26.

Each year brings headlines telling of havoc and tragedy. The best-known flood, of course, is the one recorded in *Genesis*. In the 1920's, archeologists digging near the Euphrates River in present-day Iraq unearthed royal tombs containing jeweled artifacts and furniture.

In 1929 they suddenly hit a stratum of clay deposits that, they thought, could only have been left by a flood of enormous proportions. The stratum yielded nothing. The men dug deeper and deeper, finding not even a potsherd. Finally, after digging about ten feet, they again found traces of human activity. They interpreted this as evidence of the Biblical flood, when "the waters prevailed exceedingly upon the earth." Though some archeologists still question the meaning of this finding, most believe the barren layers probably resulted from a dramatic change in the meandering course of the Euphrates River.

Hydrologists—those scientists who study the behavior of water—have determined that even a small increase in the flow of a river means a significant increase in the sediment it can carry. So a river in flood carries thousands of times more sand and silt than under normal conditions. In the lowlands, where the current slows,

"We had to shovel a foot of mud from the basement," said Robert Wood after helping clean out his mother's home in Johnstown, Pennsylvania. The tributaries of the Conemaugh River flooded the city in July 1977. In 1889, the year of the infamous Johnstown flood, some 2,200 people died after a dam burst upstream.

Swept 200 yards downstream from its foundation, a house rests on the wreckage of a bridge across the swollen Big Thompson River in the mountains of the Front Range in Colorado. A 20-foot wall of water tore through Big Thompson Canyon during a ten-inch rain on the evening of July 31,

170

TIM WILEY

MATTHEWS WESTERN COLORS (ABOVE). JOHN ROLD, COLORADO GEOLOGICAL SURVEY

1976. Before the flood, U.S. Highway 34 followed the peaceful mountain stream through the canyon; a water diversion pipe crossed overhead. After the flood, the pipe and much of the road had vanished. The flash flood struck with little warning, claiming at least 139 lives and destroying dozens of homes. "I looked out and there was a trailer tumbling down the river, bobbing in the waves like a toy in a bathtub," said one resident of the canyon. "It's unbelievable," said another. "We lost 13 neighbors." All had built homes on the river's floodplain because "it was so serene and beautiful."

Wind-tossed surf smashes the coast of California below the old Battery Point lighthouse. Since early in earth's history, oceans have gradually

the water begins to drop its sediment. Some of the burden settles, adding to the alluvial plain. The rest rides with the river to the sea.

"The Mississippi, just winding its way down to the Gulf, dumps a million tons a day, on the average," said Jim Coleman, shouting in my ear above the roar of our Cessna's engine as we flew over the river and its delta. We had boarded the plane after leaving his office at Louisiana State University's Coastal Studies Institute in Baton Rouge.

"Do you realize what a million tons means? It's the equivalent of 694 trucks each dumping a ton of dirt into the sea around the clock, every minute of the day."

From 2,000 feet we could see the slow, easy meanders of the river, the great arcs formed where the water had eroded one bank and built up another. Before long, the flat vistas of water and the alluvial plain were broken by the huddle of skyscrapers and the Superdome of New Orleans. "The city is sinking," said Jim. "Parts are already below sea level. For now, dikes and pumps can keep the water out, but I predict that in 2,000 years it will be gone. It will drown as the land

cypresses, tall roseau cane, and ghostly veils of Spanish moss.

And spilling into the Gulf in a great plume that muddied the water for 20 miles was the sediment that in 6,000 years has formed a chunk of land larger than Maryland.

The sediment is a major problem for the ships and barges that ply the river around New Orleans, the second largest port in the country after New York. The Corps of Engineers maintains the navigation channels, dredging about 240 million cubic yards of sediment from the river every year—enough to cover Delaware with a layer more than one inch thick. Patrol boats constantly monitor the river, sounding the depths to determine where shoals have formed and dredging must be done.

The Corps is fighting the river on other fronts, too. It is trying to thwart the Mississippi's natural tendency to shift its channel, to build bars and islands in one place and gnaw them away at another.

In *Life on the Mississippi,* published in 1874, Mark Twain wrote about the changes he noticed when he returned to the river after 21 years: "On the river below Memphis... wagons travel now where the steamboats used to navigate." Of the area near Cairo, Illinois, he wrote: "Goose Island is all gone but a little dab, the size of a steamboat.... One of the islands formerly called the Two Sisters is gone entirely; the other, which used to lie close to the Illinois shore, is now on the Missouri side, a mile away; it is joined solidly to the shore ... but it is Illinois ground yet, and the people ... pay Illinois taxes. ... The Mississippi is a just and equitable river; it never tumbles one man's farm overboard without building a new farm just like it for that man's neighbor."

To contain the banks and keep them from eroding, the engineers have lined the outside bends of the lower river with concrete revetments. Large slabs, held together with steel cables, form great concrete corsets anchored to the bank.

Where the Mississippi bent in long, horseshoe loops, the engineers

sculpted the world's shores—leveling, shaping, rebuilding.

sinks under the weight of the tremendous load of sediment."

As we continued south, we reached the delta. The plane banked, and I could see where Balize once stood, the 18th-century port with the reputation as the wickedest spot on the river. Now there was only brown water and the glistening raw green of the marsh. Here was the bayou country, where the river drains into a maze of sluggish waterways. Soil washed from the wheat fields of Montana and the hollows of West Virginia is drifting down slowly to form land thick with bent-kneed

The Work of Water

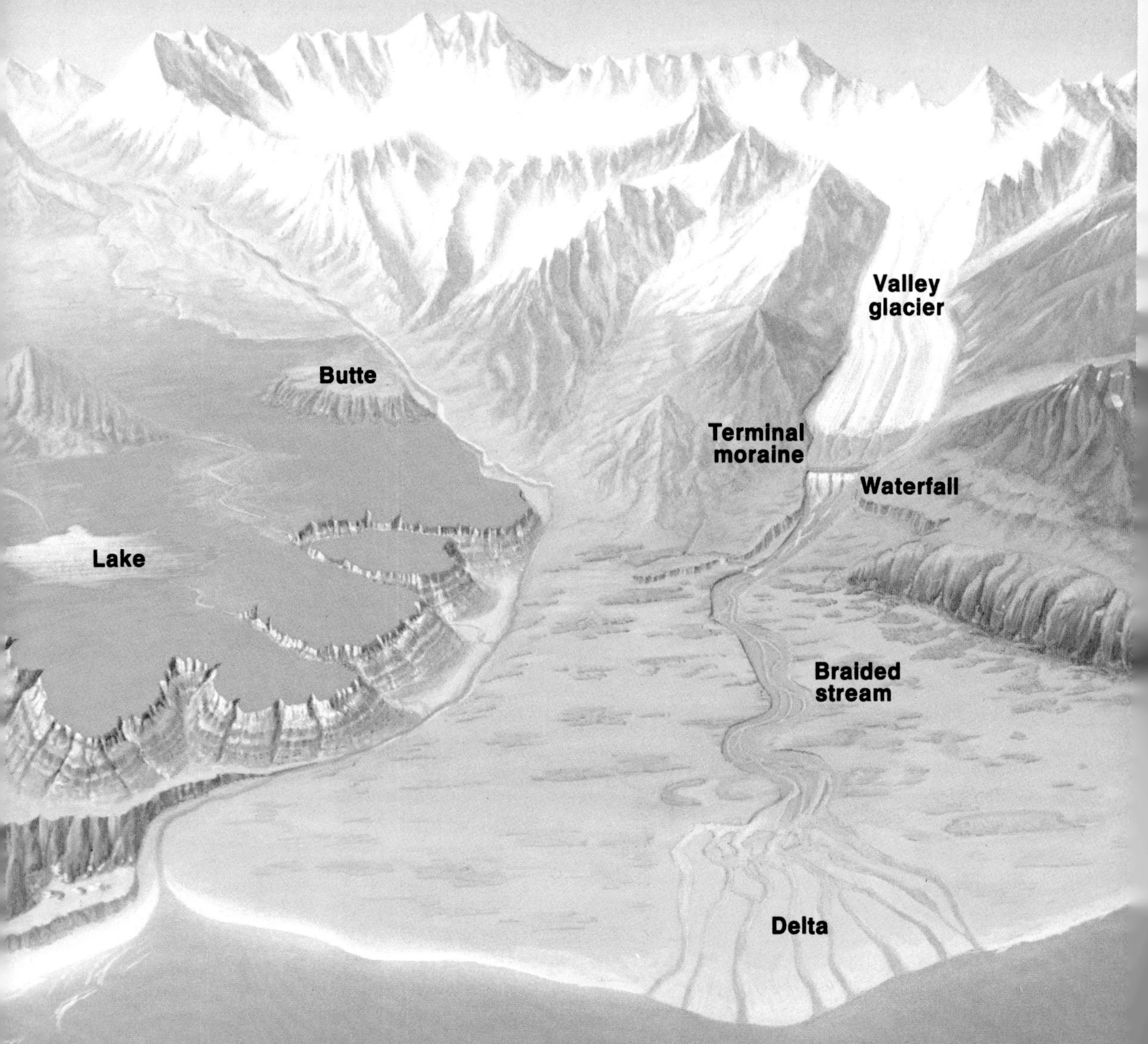

Valley glacier

Butte

Terminal moraine

Waterfall

Lake

Braided stream

Delta

18,000 years ago

Theoretical North American landscapes, separated in time by 18,000 years, reveal the ceaseless sculpturing force of water. During the last Ice Age (painting above shows representative features), heavy snow and cold temperatures in mountainous areas spawned glaciers that carved U-shaped valleys and mountainside depressions called cirques. Some of the soil and rock eroded by the glaciers came to rest in heaps and ridges called moraines; much reached the sea, carried by summer meltwaters. Today, receding glaciers have left lakes in mountain basins. Some areas that

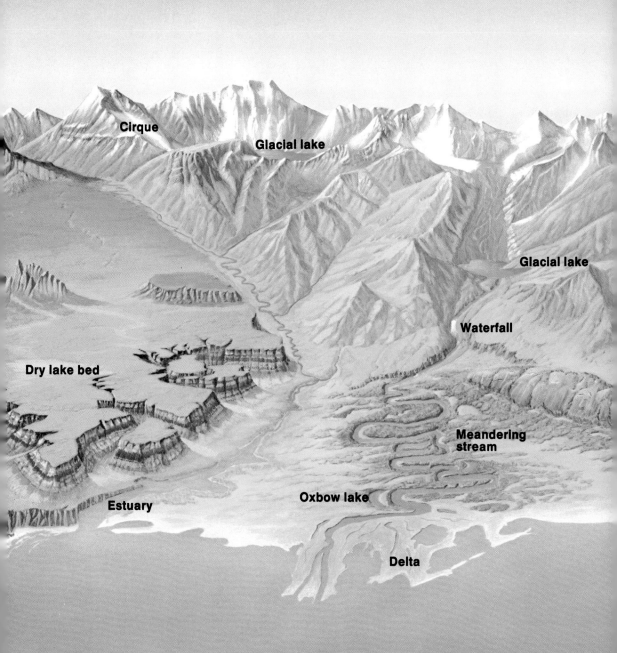

Cirque

Glacial lake

Glacial lake

Waterfall

Dry lake bed

Meandering
stream

Estuary

Oxbow lake

Delta

Today

have become desert once sup-
ported dense vegetation, and held
lakes and streams. Plants colonize
the formerly bare rock of the
mountainsides. The sea lies 400
feet higher, raised by the enormous
amounts of water released when
the Ice Age ended. Estuaries have
formed where coastal valleys

flooded. Sediment continues to
accumulate in deltas. The
meltwater river now has smaller
floods and carries less sediment. It
meanders in broad loops across the
floodplain, occasionally forming
curved oxbow lakes. Waves cut into
seaside cliffs, and a waterfall
erodes its way slowly up a valley.

PAINTING BY JAIME QUINTERO

have straightened the river by cutting channels through the narrow necks of land. Sixteen cutoffs, made in the early 1930's and '40's, lopped 151 miles from the length of the river.

Shorter and straighter, the river does not bite into its banks so severely, and floodwaters are contained between massive earthworks called levees that stretch from horizon to horizon for hundreds of miles.

Erupting in flood, the Mississippi is at once fearsome and humbling in its power. In the flood of 1927 the mile-wide river spread 80 miles in some places, and drove 700,000 people from their homes. It took between 250 and 500 lives and destroyed 200 million dollars' worth of property. The devastation moved Congress to pass the Flood Control Act of 1928 and to fund the Corps of Engineers in vast river-control projects. Has the Corps tamed the restless giant?

Charles M. Kolb, congenial geologist of the Mississippi, feels the river still has the upper hand. In the dining room of the Vicksburg Country Club, Charles shared some of his misgivings about how the river is reacting to the changes imposed on it.

"The river doesn't flood the way it used to, partly because we've had a long period of below-average rainfall, but largely because of the dams and reservoirs. Back in de Soto's time this whole valley was thick with swamps and alligators. Now people live in this valley, and it's rich. We want to keep it that way. But the reservoirs catch the flood flows, and the river has adjusted to that low flow. The river doesn't need the big channel, so it starts silting up.

"The '73 flood gave us a warning. The water crested four to six feet higher than expected. What scares me is that with the volume of water we had in the river, it shouldn't have been that high. The riverbed has silted up more than we thought, and that's dangerous — real dangerous."

Brien R. Winkley agrees. Brien, an outspoken scientist with the Army Corps of Engineers, is deeply involved in studying the river. He believes the lower Mississippi is slowly filling up and no amount of dredging can keep up with it in the long run.

He condemns the cutoffs of the '30's and '40's. "By shortening the river, we increased the gradient in those reaches. As the velocity of the water increased, it picked up more sediment. Downstream the river flattened out and dumped what it could no longer carry."

In the meantime the middle Mississippi — between St. Louis and Cairo, Illinois — is getting deeper, scouring its bed and moving more sediment downstream. And the silt keeps coming out of the hills.

"The Yazoo River and its tributaries were once nice, sinuous streams flowing into the Mississippi," Brien said. "By channeling them and straightening their banks, we've brought thousands of acres into agricultural production. But again we've made the rivers steeper. And the rivers are scouring now, affecting the whole Yazoo system. If you fly over the area you can see trees falling into creeks, banks caving in.

"Water began leveling the earth when the first raindrop fell, some three billion years ago. Here in this valley man has hastened the geologic process a thousandfold."

Hargis Hamilton, who farmed cotton and raised corn to feed his hogs, had seen it happen on his land, a few miles outside Oxford, Mississippi. At 77, Hargis is still a busy man, operating a fence post mill from the little shed beside the big pecan tree in Taylor. He was paying his help when I arrived. From a great wad of bills, he peeled off a series of tens and twenties and jammed the rest deep into his trouser pocket. Then with one last direct hit into the spittoon, he reached for his straw hat and we walked out into the glare of a Mississippi summer day.

"I plowed every foot of these hills, and now they're all bare. The hills got washed away, and the creeks filled up with sand. We done wrong ever cleaning up these hills. We should have kept them in timber. I helped Dad clear a piece of ground when I was 10. By the time I was 25 it was all gullied up. Come a big rain, it would start. A little wash down across a hill. It got bigger each year. Soon I couldn't get a mule across. Finally it got so big I could have put my house inside. After six or seven years the topsoil was all gone, and there was nothing to make anything grow."

In 1947, the Soil Conservation Service planted pine trees and a fast-growing vine called kudzu to control the erosion of Hargis's farm, but the rich soil was gone.

"Rain! whose soft architectural hands have power to cut stones, and chisel to shapes of grandeur the very mountains."

— HENRY WARD BEECHER

"Visualize each raindrop striking the earth as a little bomb," said L. Donald Meyer, leader of soil erosion research at the United States Department of Agriculture's Sedimentation Laboratory in Oxford.

"It's not the downward pressure of the drop, but the explosion of it outward that breaks the particles loose. The drops splatter like tomatoes thrown against a wall. We've computed the energy striking Mississippi in the course of a year's rainfall, and it is the equivalent of a billion tons of TNT hitting the state." In a nearby cotton field, Don and I skirted the spray of a rainfall simulator and watched the drops strike the ground and splash up the loosened soil. At first the ground absorbed the water, but little by little the soil surface began to glisten as it became saturated. It wasn't long before the excess began flowing off, carrying soil with it.

"As the drops bombard the soil, they build a surface seal which reduces water intake and causes more runoff. The increasing flow moving down a slope can really tear things up and erode rills and gullies. Clay soil is cohesive and doesn't erode unless subjected to considerable energy. Silty and sandy soils break apart much more quickly. The steepness and length of slope are factors, too. And soil that is protected by vegetation is much less erodible than soil that is bare."

"But," said Alan Howard, youthful geomorphologist at the University of Virginia in Charlottesville, "weathering — not just rainfall — is the start of the erosion process.

"Geomorphologists," he told me, "try to understand the origin of landscapes — how they are formed and shaped." Outdoors, the Virginia city sparkled with the vigor of spring. Tulip trees gleamed with fresh, new leaves. The dogwoods were flowering, each tree a canopy of pink or white, and the lush and radiant azaleas flamed with color.

Inside, in Alan's book-lined office, I was learning how water, the primary tool of weathering, breaks down bedrock into small pieces. It dissolves rocks like so much sugar or crumbles them into fragments that become the ingredients of soil.

Acid-bearing water, which has picked up carbon dioxide from decaying plants in the ground and from the atmosphere, dissolves limestone and carries it off in solution. "You're unaware of it until it turns up deposited inside your teakettle or water tank," said Alan. "Or unless you descend into the chambers leached out of limestone in such places as Mammoth Cave in Kentucky and Carlsbad Caverns in New Mexico."

Like limestone, feldspar reacts chemically with water. One of the most common minerals in rock, it weathers to grains of clay that can easily wash or blow away or accumulate to help form soil.

Alan then talked about water as a mechanical force that causes rock to disintegrate. "Water, as it freezes and thaws in a chink or crevice, exerts a tremendous pressure that eventually splits the rock apart."

Everyone knows that water expands when it changes to ice. But I had not realized its crushing power. In becoming ice, the water swells 9 percent over its original volume. It forms a wedge that may exert the astonishing force of 30,000 pounds per square inch.

(Continued on page 186)

Havasu Falls, swollen by rain and tinted by sediment eroded from rocks upstream, plunges a hundred feet in Arizona's Grand Canyon. Calcium deposits from springs can build the falls half an inch higher each year; rare floods cut it back down.

Water—in rushing waterfalls or single drops—chisels the land. A raindrop magnified eight times gouges a hole and scatters soil as it bursts on impact. Millions of drops pummel an acre of land every second during an average rainfall. By weathering minerals from rocks, water helps build fertile soil. In time, it carves gorges and levels mountains, wearing away even the hardest rock.

Boatmen challenge a stretch of white water on the Colorado River, often fast and turbulent as it surges through the Grand Canyon. Half a billion years ago a broad plain stretched across this region. On it, ancient seas, rivers, and wind-blown sand deposits built up sediment thousands of feet thick. At least ten million years ago, geologic forces began uplifting the plain, creating the Colorado Plateau. The river cut downward, carving the mile-deep gorge, and continues to do so. Tributaries have eroded side canyons such as Deer Creek (below, right). Its layer-cake walls record some 500 million years of earth's history. At right, along the Canyon's North Rim, rock layers hold marine fossils that document ancient invasions by the sea.

Jetties reach rocky fingers
into the Mississippi River,
trapping sediment and
helping to keep the river's channel
clear for shipping. The
Mississippi River system,
draining water from 41 percent
of the contiguous United States,
carries millions of tons of sediment
each year to the Gulf of Mexico,
forming a delta of some 12,000
square miles. "False-color"
satellite photography provides a
portrait of the delta: Light blue
indicates sediment-laden waters;
red shows vegetation. Below, a
crewman pries loose a log caught
in the cutter head of a dredge. The
U. S. Army Corps of Engineers
works the river from mid-
February to October to ensure
a navigable channel.

NASA (ABOVE); NATHAN BENN (BELOW); N.G.S. PHOTOGRAPHER JAMES L. STANFIELD (OPPOSITE)

Preceding pages: Concrete slabs,
strung together with steel cables,
shield a bank of the Mississippi
River near Reserve, Louisiana.
Engineers have laid miles of such
revetments to help protect the
river's banks from erosive currents.

PRECEDING PAGES: NATIONAL GEOGRAPHIC
PHOTOGRAPHER JAMES L. STANFIELD

Once, while camped above tree-line, I heard the sound of frost at work. Again and again, the clatter of rock ricocheting against rock pierced the silence. The boulders rolled and bounded downhill until they came to a halt on the talus slope that fanned out from the base of the cliff.

All day the sun beat down, dazzling and warm. Then, when the sun faded from a pink sky, dew and moisture turned to ice. All around me in the nocturnal chill, ice was prying rock apart with mute blows that would wear these splendid San Juan Mountains in Colorado to blunt nubs.

"The same action of swelling and shrinking plays a part in the slow movement of soil downhill," said Alan. "We call it soil creep." At each frost or rain, the ground expands, and then with a thaw or dry spell it contracts. The heaving and buckling dislodge particles of soil or weathered rock, and the force of gravity moves them slowly downhill.

"It's not noticeable from year to year, but you can see the effects on hillsides where poles, fence posts, stone walls, and gravestones tilt downhill as the soil around them moves. In the long run, we believe, creep moves much more material, particle by particle, than landslides."

Soil creep may be barely perceptible, but land sliding downhill can strike with terrifying force. About a hundred lives were lost in 1969 when heavy rains sent thousands of tons of earth plummeting down the San Gabriel Mountains rimming Los Angeles. In Italy and Peru in the past 20 years, soil saturated with water has broken loose from slopes, causing massive slides that snuffed out thousands of lives.

"Water, in one way or another, not earthquakes, has been responsible for most landslides," said Bruce M. Kaliser of the Utah Geological and Mineral Survey. "Most landslides are touched off by heavy rains. Sometimes the bedrock gives way under the added weight and pressure of water. Sometimes it's the soils that are unstable. Clay, for instance, becomes slippery when wet.

"And man, by cutting into hill-

sides and altering the vegetation, contributes to the problem, too. For example, some stream banks were cleared of vegetation and brush and reseeded in grass to get more water flowing to Los Angeles. The larger plants, it was said, absorbed too much water. Then when the rains came in 1969, the root systems were not there to hold the soil.

"Landslides are more frequent than people realize. Did you know," he asked, "that the building of the Panama Canal caused a plague of landslides? The day after it was

opened, enough rain-drenched jungle slumped into the canal to close it for days. Dredging the locks and propping up the slopes are still full time jobs costing millions of dollars a year."

Not far from Mr. Kaliser's office — and at all of Salt Lake City's backdoor — loom the lofty peaks of the Wasatch Range. There the high treeless slopes and imposing ridges harbor thousands of potential landslides. In winter it is snow that hurtles down the steep mountains, crackling and hissing in the deep stillness.

Cascade of snow thunders down a forested slope of the Columbia Mountains near Canada's Glacier National Park. Avalanches can exceed 100 miles an hour, snapping trees or uprooting them, stripping hillsides and exposing them to erosion. In some areas, patrols use explosives to trigger snowslides, reducing the risk of major avalanches.

River of ice, Taku Glacier advances some 500 feet annually as it spills from the Juneau Icefield, a vast sheet of ice 1,700 feet thick in places. Members of a field team measure its rate of movement. Taku flows through steep canyons in the coastal mountains, its surface buckling and fracturing as it moves. More than 100 feet of snowfall feed the crevasse-scarred glacier each year. Near its tip (right), meltwater piles up silt, sand, and fragmented debris before flowing into Taku Inlet. Advancing and retreating, massive ice sheets have sculpted much of the earth's surface, carving spectacular valleys and ridges.

"White death" is what the people in the Alps call the snow avalanches that through the centuries have crashed down on their towns and villages. During the First World War thousands of Austrian and Italian troops in the Alps were victims not of shells and bullets but of snow. In the Tyrolean campaign, an avalanche demolished and buried an entire barracks in a single slide.

Most avalanches occur on steep slopes during or just after big storms when the new snow is too heavy to stick. I learned some of the things that lead to avalanches from Don Bachman, an avalanche forecaster with the U. S. Forest Service.

"The odds favor an avalanche when a heavy snowfall rests on an icy surface. Snow falling on a fragile snowpack, especially when accompanied by strong winds, also sets the stage for avalanche release. In the spring, thaw conditions will lead to wet snowslides. Most avalanches start naturally, but many people trigger the very slide that buries them. Any time layers of snow perch on top of each other, a skier swooping downhill can trigger an avalanche."

Michael Lahey, a member of the National Ski Patrol, witnessed an avalanche in the Skyline Ski Area in Idaho and wrote about it in the magazine *Skiing*. "There was a low rumbling, like a growl... A tremendous boom sounded, followed by a fracture line just five feet below my ski tips. The mountain below me disappeared, and the snapping aspen branches were mixed with orange-jacketed patrolmen who were swept away by the tidal wave of snow."

When the slide stopped, Michael and the other skiers in the group probed the area with ten-foot rods used in avalanche rescue. Two of the victims were buried and died of suffocation. The 14 who survived had been only partially buried.

Countless snowflakes and centuries of cold create another formidable force that has chiseled, scoured, and planed the landscape of more than half of North America. When snow accumulates more rapidly than it melts, when it piles up year after

year, the lower layers are compressed; under this increasing pressure, combined with the altering effects of refreezing, the flakes recrystallize into ice. As the mass grows thicker and larger, its own tremendous weight pushes the ice downward from the high, cold mountains to the lower valleys. When it starts to move, the ice has become a glacier.

It is a river of ice sliding very slowly, so slowly that I was unaware

---

"Nothing on earth is so weak and yielding as water, but for breaking down the firm and strong it has no equal."

—LAO-TZU, *Chinese Philosopher, 6th century B.C.*

---

of it. It was July, and I was skiing the Ptarmigan Glacier, trying to follow Maynard Miller, who kept disappearing in the blowing mist that swirled around us. Some 4,000 feet below, hidden behind a bank of clouds, lay Juneau, Alaska. Stretching behind us, to the north and east, were nearly 2,000 square miles of perpetual snow and ice, called the Juneau Icefield—1,700 feet thick in places and nourishing more than three dozen major glaciers and hundreds of lesser ones.

Dr. Miller has studied glaciers in Alaska since 1940 and directed the Juneau Icefield Research Program since 1946. That morning he had described how glacial ice is formed.

There was so much to learn— how glaciers scrape and scour their beds as they move; why they advance and recede. But we were growing restless indoors. Glaciers were honing their knife-edge ridges, polishing rock, and plowing up moraines on the valley walls right outside the door. A field trip was decreed.

We bent against the wind as we stepped into our skis and started downhill for the $2^1/_2$-mile descent to the ice terminus, where the glacier deposits its burden of rock and debris.

The fog swallowed the other skiers, and I seemed to be alone. There was only ice and the rock flanks of mountains rising out of the snow and looming darkly above me. I could have been back in one of the ice ages, when glaciers gripped the continent as far south as Kentucky.

"It's so quiet you can hear the blood in your head," said Lee Schoen, the brawny and capable deputy leader who swooshed out of the fog to halt beside me.

"Look," he said, pointing to red specks in the snow.

"Algae, here!" It was hard for me to believe.

"Sure, and look closely," he said, tugging at my poncho. I crouched and saw tiny black specks, like fine pepper, in motion.

"They're snow fleas," Lee said. "*Collembola,* or springtails—a tough little animal. They live on the ice and feed on the algae."

At the terminus of the glacier, where the group was waiting, we shed our skis and continued on foot. We walked on the rock debris that had been scraped from the land, brought down the valley, and released as the glacier melted at its margin. "The Ptarmigan Glacier is a receding one," Maynard Miller told me as we hiked slowly downhill. "The snowfields that nourish it are just not at high enough an elevation to survive the melting effects of our warmer summers."

In the distance of half a mile we walked from rubble left recently to that deposited in the middle of the 18th century. We went from a barren realm to where rock was embossed with black and green lichens—stubbornly alive on the cold rock. Then we came to clumps of moss bordering the moraine. A little beyond them the moss spread into a springy carpet of green heather. We had arrived at the tundra, where tiny dwarf flowers created a pageant of color—yellow arctic buttercups, purple saxifrage, deepblue lupines.

In ten minutes I had seen in microcosm the earth emerging from an ice age.

When a glacier advances at the same rate as it melts, the debris piles up at the toe in a ridge of unsorted soil and rock called a terminal moraine. Such moraines were built by the ice sheets that covered much of the northern tier of states from New England to Washington during the Wisconsin Glacial Age, the last of four major ice ages that occurred during the last million and a half years. When the ice retreated, it left loose sand, gravel, and clay, called till. Long Island, Nantucket, and Cape Cod were built of such materials. As the melting ice lost its grip, it also dropped a cargo of boulders carried from as far away as Hudson Bay and Labrador. Plymouth Rock itself is a large glacial boulder. Others are the small stones that plague farmers from Massachusetts to Michigan. And in some places, where bedrock was soft, or where there was delayed melting of huge ice blocks, lakes and ponds were formed. The Connecticut Valley itself was once a glacial lake; it drained just 11,000 years ago. Glaciers carved resistant bedrock, too. Yosemite Valley was gouged from hard granite.

No one could ski or walk on the last few miles of Taku Glacier, one of Alaska's healthiest advancing glaciers. It's a jumble of ice slashed with crevasses, formed wherever the ice has moved at an uneven rate over sloping bedrock or around corners in the valley. The gashes expose the dazzling blue of glacial ice that gleams like cold fire from the vault of a hundred winters.

"If you compressed snow enough, it would be blue, too," said Lee. "Snow and ordinary ice are white because there's a lot of air in them."

We were looking down on the brutal expanse of the glacier from a helicopter. Although Maynard Miller had explained that the Taku was an advancing glacier "racing" into the rain forest at the rate of about a foot and a half a day, I was not prepared for the destruction I saw beneath us.

We whirred down the icefield to the glacier's edge, where we followed for two miles its frontal perimeter, littered with broken and splintered trees. Summer's thaw had exposed last winter's invasion, when the glacier had rammed into spruce and hemlock 200 years old. In the coming winter the glacier would roll over them again and continue its steady advance. Maynard believes that within a decade the glacier will cross the Taku Inlet and block this access to the sea. Eventually the glacier will recede—as it has every 200 years—and leave a paradise of sharply etched mountains. In time the mountains will weather and erode and the resulting sediments settle to the bottom of the sea, building new land along this edge of the continent.

A newcomer to the science of geology, I had in my travels gained some insight into how water and time transfigure the landscape. I had seen places where they had endowed the earth with fertile valleys and memorable landscapes. Water and time had formed the silhouette of the shore where I swim, molded the gentle contours of northern Virginia where I live, and created the sedimentary rocks that mantle three-quarters of the earth's land surface.

And there is a place that turns time into rock and takes you back through almost two billion years of earth's history, a footstep at a time.

I took that journey down into the Grand Canyon with George Billingsley, a young geologist who has hiked in the canyon more than 100 times and never wearies of its wonders. We gazed in silence at the immensity of a planet opening its heart. "Each foot of vertical descent from the rim to the bottom takes us back 400,000 years," he said as we started down the Kaibab Trail toward the Colorado River, which was almost a vertical mile down and seven miles away.

We descended through the alternating layers of limestone, sandstone, and shale. We traversed about

Taku's summer thaw exposes winter's destruction. The advancing glacier splinters 200-year-old trees and builds up a lateral moraine of rocks and forest debris. Scientists check the extent of the summer melt. At left, geologist Dr. Maynard Miller explores a 60-foot-deep crevasse. "It's wet, eerie, cold, and clammy—but incredibly beautiful," he says. Stress caused by uneven rates of glacial movement creates the deep crevasses.

half a billion years. Ten seas had come and gone. The shells of untold billions of snails, clams, and other creatures had disintegrated to help build limestone beds hundreds of feet thick. Great river systems had surged and withered, depositing mud and sand to form shales and sandstones.

Finally we reached the sheer walls of the Inner Gorge, and on a trail of switchbacks cut into rock called Vishnu Schist we began to hike the last 1,200 feet to the river.

The air quivered with the heat. George looked at his pocket thermometer. "A hundred twenty degrees. You can count on it getting five degrees hotter for every 1,000 feet we descend," he said. I ran my hand over the dark gray Vishnu Schist. "It was formed from the debris of mountains flattened by millions of centuries of erosion," George told me.

If ever there's a place, I thought, where we can penetrate the long dark shadows of geologic time, it is here at the bottom of the Grand Canyon.

I had been prepared for the spectacle of the canyon — the awesome chasm where water had carved a vast and beautiful network of gorges and gashes in the earth's surface. But the wonder lies not in the great voids or in the countless fragments that have been flaked away to form the rocky abyss. Rather it arises from the realization that all the cliffs and terraces, all the ramparts of stone, all the eroded spires and buttes, all the tiers of rock, were built of layers of sediment accumulating one day at a time.

At the bottom of the canyon I stood in awe of what water and the ages had created.

Lava cliffs rise along the coast of Kauai in Hawaii. Millions of years ago ocean waves began carving cliffs from the volcanic rock; rainfall has eroded deep gullies. Where nature's powers create new land, water ceaselessly alters it. No surface escapes the irresistible force of rushing streams, grinding glaciers, pounding waves — or falling rain.

Notes on Authors

WILLIAM R. GRAY, a native of Washington, D. C., and a graduate of Bucknell University, joined the Society's staff in 1968. A writer and editor — and an avid outdoorsman — he is author of *The Pacific Crest Trail,* a Special Publication, and *Camping Adventure,* one of the Society's Books for Young Explorers. He has contributed chapters to *Alaska: High Roads to Adventure* and *The Alps.*

A journalist who received her bachelor's degree from the University of Missouri, and a master's degree from The American University, TEE LOFTIN joined the Society's staff in 1967. Author of the Special Publication *The Wild Shores: America's Beginnings,* she has also contributed chapters to *Those Inventive Americans* and *Clues to America's Past.* She was born near Kinston, North Carolina.

TOM MELHAM, born in Milwaukee, grew up on Long Island, New York. He received a bachelor's degree in biochemstry from Cornell University and a master's degree in journalism from the University of Missouri. Author of the Special Publication *John Muir's Wild America* and a contributor to *The Craftsman in America,* he has been on the Society's staff since 1971.

Managing editor of the Society's Books for Young Explorers, a series of volumes for children between the ages of four and eight, CYNTHIA RUSS RAMSAY is a native of New York City and a graduate of Hunter College. Since joining the Society's staff in 1966 she has contributed to several Special Publications, including *The Alps, Life in Rural America,* and *Alaska: High Roads to Adventure.*

JUDITH E. RINARD has been a writer of National Geographic Educational Filmstrips since 1972. She has also written two of the Society's children's books — *Wonders of the Desert World* and *Creatures of the Night.* She was born and raised in Iowa, and received her bachelor's degree in English from the University of Toronto.

Acknowledgments

The Special Publications Division is grateful to the individuals, agencies, and organizations named or quoted in this book and to those cited here for their generous cooperation and help during its preparation. Consultants: Dr. Robert W. Decker, Professor of Geology, Dartmouth College; Dr. Alan D. Howard, Professor of Geomorphology, University of Virginia; Patrick E. Hughes, Chief, Publications and Media Staff, NOAA; Douglas M. LeComte, Meteorologist, NOAA; Dr. J. Murray Mitchell, Senior Research Climatologist, NOAA; Malcolm Reid, Meteorologist, Environmental Data Service, NOAA; Dr. Peter B. Stifel, Professor of Geology, University of Maryland; Thomas A. Wilson, Meteorologist, NOAA; Dr. Augustine Y. M. Yao, Research Meteorologist, NOAA. Robert F. Abbey, Jr., Dr. William Breed, Dr. Richard S. Fiske, Henry J. Frentz, Dr. Robert M. Hamilton, John R. Hope, Dr. Edwin Kessler, H. Michael Mogil, Vincent J. Oliver, Dr. Donald W. Peterson, and Dr. Robert L. Wesson. Museum of Northern Arizona, USGS Hawaiian Volcano Observatory.

Additional Reading

The reader may wish to consult the *National Geographic Index* for related articles, and to refer to the following books: Bruce W. Atkinson, *This Weather Business;* Louis Batton, *The Nature of Violent Storms;* B. A. Bolt, W. L. Horn, G. A. Macdonald and R. F. Scott, *Geological Hazards;* Fred Bullard, *Volcanoes of the Earth;* T. J. Chandler, *The Air Around Us;* James Cornell, *Great International Disaster Book;* Pete Daniel, *Deep'n As It Come: The 1927 Mississippi River Flood;* Peter Francis, *Volcanoes;* Sir Vivian Fuchs, ed., *Forces of Nature;* William Glen, *Continental Drift and Plate Tectonics;* Stephen Harris, *Fire and Ice;* Patrick Hughes, *American Weather Stories;* Gordon Macdonald, *Volcanoes;* Charles F. Richter, *Elementary Seismology;* Elmer R. Reiter, *Jet Streams: How Do They Affect Our Weather;* Stephen Schneider, *The Genesis Strategy;* Walter Sullivan, *Continents in Motion;* Dorothy B. Vitaliano, *Legends of the Earth;* Peter Yanev, *Peace of Mind in Earthquake Country.*

Composition for POWERS OF NATURE by National Geographic's Photographic Services, Carl M. Shrader, Chief; Lawrence F. Ludwig, Assistant Chief. Printed and bound by Kingsport Press, Kingsport, Tenn. Color separations by Colorgraphics, Inc., Forestville, Md.; Graphic South, Charlotte, N.C.; National Bickford Graphics, Inc., Providence, R.I.; Progressive Color Corp., Rockville, Md.; J. Wm. Reed Co., Alexandria, Va.

Library of Congress ⊂IP Data

National Geographic Society, Washington, D. C., Special Publications Division. Powers of Nature.

Bibliography: p. 197
Includes index.
1. Earthquakes.   2. Volcanoes.
3. Meteorology.   4. Droughts.
5. Water. I. Title
QE33.N29   1978   551   76-57002
ISBN 0-87044-234-1